BRIEF HISTORY OF HERPETOLOGY IN THE MUSEUM OF VERTEBRATE
ZOOLOGY, UNIVERSITY OF CALIFORNIA, BERKELEY, WITH A LIST OF
TYPE SPECIMENS OF RECENT AMPHIBIANS AND REPTILES

Robert Cyril "Bob" Stebbins examining a specimen of the salamander *Ensatina eschscholtzii* in a redwood grove, Berkeley campus, in 1951. Photograph by Oliver P. Pearson.

Brief History of Herpetology in the Museum of Vertebrate Zoology, University of California, Berkeley, with a List of Type Specimens of Recent Amphibians and Reptiles

Javier A. Rodríguez-Robles,
David A. Good, and
David B. Wake

A Contribution from the Museum of Vertebrate Zoology
of the University of California, Berkeley

University of California Press
Berkeley · Los Angeles · London

UNIVERSITY OF CALIFORNIA PUBLICATIONS IN ZOOLOGY
Editorial Board: Peter Moyle, James L. Patton, Donald C. Potts, David S. Woodruff

Volume 131

UNIVERSITY OF CALIFORNIA PRESS
BERKELEY AND LOS ANGELES, CALIFORNIA

UNIVERSITY OF CALIFORNIA PRESS, LTD.
LONDON, ENGLAND

Library of Congress Cataloging-in-Publication Data
Rodríguez-Robles, Javier, 1964–, David A. Good, 1956–, and David B. Wake, 1936–
    Brief history of herpetology in the Museum of Vertebrate Zoology, University
of California, Berkeley, with a list of type specimens of recent amphibians
and reptiles / Javier Rodríguez-Robles, David A. Good, and David B. Wake
    p. cm. (University of California publications in zoology ; 131)
"A contribution from the Museum of Vertebrate Zoology of the University of
California, Berkeley. "
Includes bibliographical references.
ISBN 0-520-23818-4 (pbk. : alk. paper)
    1. Reptiles—Type specimens—Catalogs and collections—California—Berkeley.
    2. Amphibians—Type specimens—Catalogs and collections California—Berkeley.
    3. Herpetology—California—Berkeley—History. 4. University of California,
Berkeley. Museum of Vertebrate Zoology.

QL645.2 .R64 2003
597.9/074/79467—dc21   2002067588

Dedication

We dedicate this volume to our colleague and friend, Robert Cyril Stebbins, affectionately known as Bob, who established the modern herpetological tradition in the Museum of Vertebrate Zoology (MVZ) in 1945 and who remains actively involved in herpetology today. Bob was born in Chico, Butte County, California, on March 31, 1915. His family moved to the San Fernando Valley, near Los Angeles, where Bob wandered the hills and valleys and, following in his father's footsteps, became a naturalist. Bob attended North Hollywood High School and received his B.A. (1940, with highest honors), M.A. (1942), and Ph.D. (1943) from the University of California, Los Angeles. He attended the Yosemite Field School of Natural History in summer 1940 and served as a Ranger-Naturalist at Lassen Volcanic Park during the summers of 1941 and 1942. Bob joined the University of California, Berkeley, faculty in 1945 and retired as Curator Emeritus of Herpetology in the MVZ and Professor Emeritus of Zoology in 1978.

As a graduate student, Bob was particularly interested in the physiological ecology of desert lizards, but he had begun ecological studies of salamanders in the Santa Monica Mountains near his home. When he moved to Berkeley he continued to work with salamanders, and his detailed studies led to two monographs on the salamander *Ensatina eschscholtzii* (Stebbins, 1949, 1954b), which have proven his most enduring research contributions. His argument that *Ensatina* is a ring species attracted a great deal of attention and has continued to be a focal point for studies not only of the complex itself but also of the more general issue of the ring species idea. When Bob was establishing his research and education program at Berkeley, he sponsored several graduate and postdoctoral students who also conducted research on amphibians, among them, John R. Hendrickson Jr., Charles H. Lowe Jr., William J. Riemer, Gerson M. Rosenthal, and Frederick Turner.

Once settled in Berkeley, Bob began a series of extensive field studies that resulted in his first book, the substantial and still valuable *Amphibians of Western North America* (1951). He later condensed the amphibian accounts and added reptilian accounts to produce the widely used *Amphibians and Reptiles of Western North America* (1954a). As Curator of a growing collection, he undertook systematic studies (e.g., Stebbins, 1958a; Stebbins and Lowe, 1949, 1951), especially early in his career. However, his primary interest was the biology of living organisms. Throughout his career, Bob considered himself a natural historian. His contributions ranged broadly from his early work on the adaptive morphology of the nose of fringe-toed lizards (*Uma* spp.; Stebbins, 1943), adaptive coloration in animals (Ipsen et al., 1966a, b; Stebbins et al., 1966), antipredator behavioral responses of organisms (Leonard and Stebbins, 1999), and, notably, the biological role of the parietal eye (the "third eye") of tuataras (Stebbins, 1958b) and lizards (e.g., Eakin and Stebbins, 1959; Stebbins, 1970; Stebbins and Eakin, 1958). Whereas he concentrated mainly on western North America, on sabbatical leaves he studied the natural history of amphibians and reptiles in Colombia (Stebbins and Hendrickson, 1959), South Africa (Stebbins, 1961; Stebbins et al., 1960), the Galápagos Islands (Stebbins et al., 1967), and Australia (Stebbins and Barwick, 1968). He is an avid note taker, as evidenced by the 35 bound volumes of his field notes, taken from 1945 through 1990, that are on file in the MVZ, with more to come. His detailed notes are illustrated with his exceptional and sometimes hilarious drawings (Fig. 1).

Bob feels strongly that education is the only way to solve environmental and population problems, and he has effectively conveyed this message to students, colleagues, and the general public. He developed an interest in elementary and secondary school science education and, among other numerous activities, served as Chairman of the University of California Elementary School Science Committee from 1960 to 1961, participated in a National Academy of Sciences–sponsored program devoted to introducing guidelines for laboratory science in secondary schools in various Asian countries in 1963, and produced two educational films for the Sierra Club. Bob's devotion to organisms and their surroundings also led him into environmental activism. A special area of concern for him was the California deserts, and he played a significant role in regulating the use of off–road vehicles in these fragile environments (Stebbins, 1974a, b). In 1995 Bob received the Environmental Spirit Award "in recognition of his extraordinary contributions to science, and for his quiet, unassuming ability to inspire respect for the natural world."

From the beginning of his career, Bob illustrated his research publications with his own drawings. His several important books brought him wide recognition and acclaim, especially his *A Field Guide to Western Reptiles and Amphibians*, first published in 1966 (and revised in 1985) in the Peterson Field Guide series. It is noteworthy that Bob's original passion was birds, and with his father, Cyril A. Stebbins, he had written two field guides to birds of California (Stebbins and Stebbins, 1953, 1963). When Bob retired, he turned to animal portraiture and landscape painting; several of his superb canvases are on permanent display in the MVZ and in the Marian Koshland Bioscience and Natural Resources Library on the Berkeley campus. However, he remained active in herpetology and in 1995 published another book, *A Natural History of Amphibians*, with his longtime friend and associate, Nathan W. Cohen. At present Bob is completing the third edition of his popular field guide, for which he has prepared many new color paintings, among his best. This extraordinary scientist remains an active and valued member of the MVZ community of scholars.

Contents

Figures and Tables

Acknowledgments

We thank Aaron M. Bauer, Charles J. Cole, Alan de Queiroz, Harry W. Greene, Peter A. Holm, John B. Iverson, Joseph R. Mendelson III, James F. Parham, Tod W. Reeder, Hobart M. Smith, and Van Wallach for clarifying amphibian and reptilian taxonomy; Harry W. Greene and Theodore J. Papenfuss for encouragement and discussions; Roy W. McDiarmid, Christopher J. Bell, Barbara R. Stein, and James L. Patton for valuable comments on earlier versions of the manuscript; and Robert C. Stebbins, William Z. Lidicker Jr., John R. Wieczorek, Elizabeth J. Holland, E. Anne Caulfield, Lorna Fay, Carla Cicero, Millicent Morris-Chaney, Joan Lubenow, Luis F. García, Kraig Adler, Dale R. McCullough, Alex Yu, Francisco F. Pedroche, and Amy E. Jess for information and assistance.

Abstract

An overview of the herpetological program of the Museum of Vertebrate Zoology (MVZ), University of California, Berkeley, is presented. The history of herpetological activities in the MVZ and more generally at Berkeley is summarized. Although the MVZ has existed since 1908, until 1945 there was no formal curator for the collection of amphibians and nonavian reptiles. Since that time Robert C. Stebbins, David B. Wake, Harry W. Greene, Javier A. Rodríguez-Robles (in an interim capacity), and Craig Moritz have served in that position. All type specimens of recent amphibians and nonavian reptiles in the collection are listed. The 1,765 type specimens in the MVZ comprise 120 holotypes, three neotypes, three syntypes, and 1,639 paratopotypes and paratypes; 83 of the holotypes were originally described as full species. Of the 196 amphibian and nonavian reptilian taxa represented by type material, most were collected in México (63) and California (USA, 54). Information is also provided concerning the collection. The first entry in the herpetological catalog in the MVZ was made on March 13, 1909 (MVZ 1, *Crotaphytus bicinctores*), and as of December 31, 2001, the collection contained 232,254 specimens of amphibians and nonavian reptiles. Taxonomically, the collection is strongest in salamanders, accounting for 99,176 specimens, followed by "lizards" (squamate reptiles other than snakes and amphisbaenians) (63,439), frogs (40,563), snakes (24,937), turtles (2,643), caecilians (979), amphisbaenians (451), crocodilians (63), and tuataras (3). A list, exclusive of abstracts and book reviews, of more than 1,300 articles, book chapters, and books on any aspect of the biology of amphibians published by researchers associated with the MVZ since its founding is available online at http://www.mip.berkeley.edu/mvz/collections/MVZHerpPubs.htm.

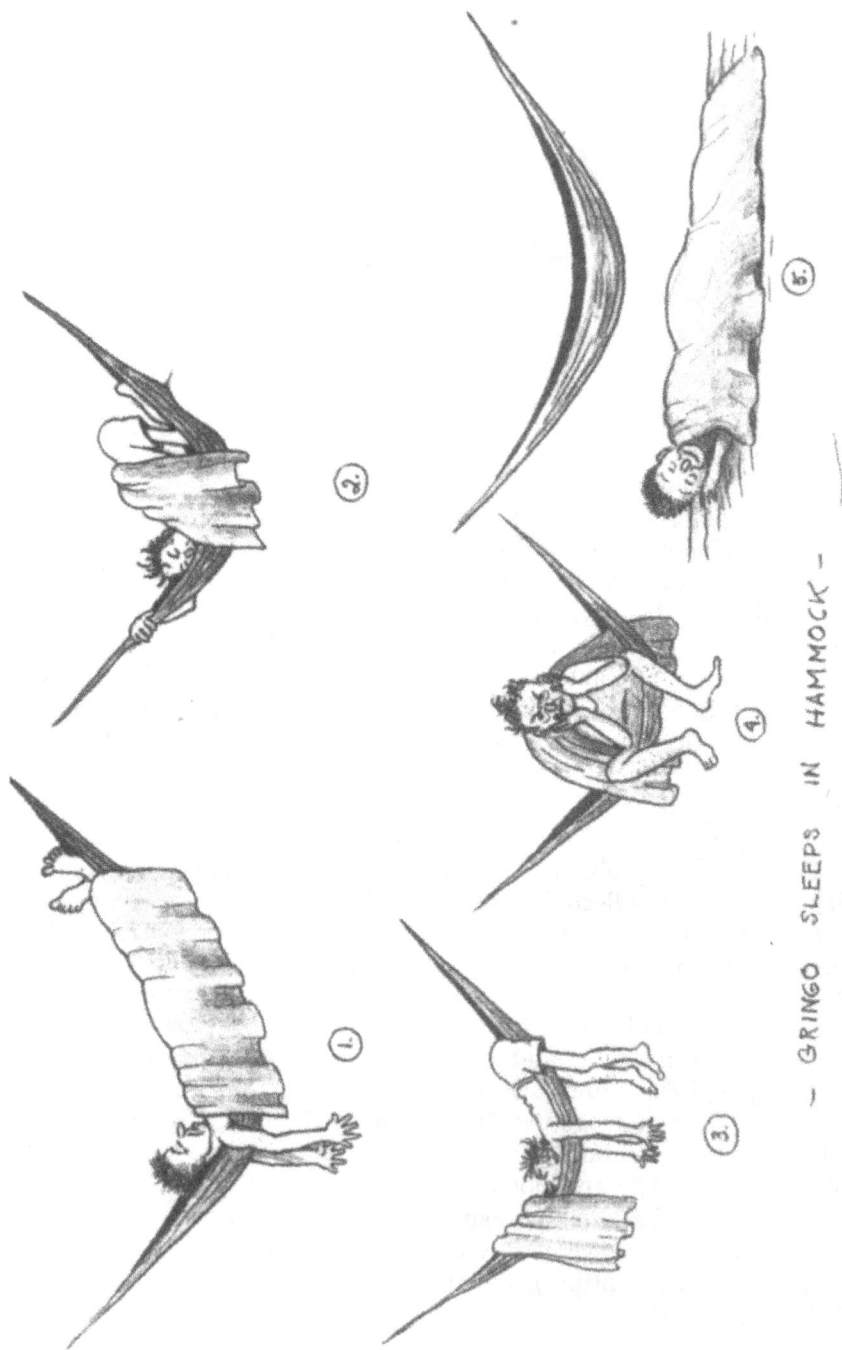

Figure 1. "Tonight I slept, for the first time, in a hammock. I may get used to it. My first night was fitful." Robert Stebbins, Sierra de la Macarena, Departamento del Meta, Colombia, November 25, 1950. (Taken from "FIELD NOTES. Stebbins. R. C. 1950, P.T.," filed in the Museum of Vertebrate Zoology.)

BRIEF HISTORY OF HERPETOLOGY IN THE MUSEUM OF VERTEBRATE ZOOLOGY

The Museum of Vertebrate Zoology, located on the campus of the University of California, Berkeley, is a leading center of herpetological research in the United States. We offer a brief account of the principal figures associated with the collection and of the most important events in the history of herpetology in the MVZ during its first 93 years. This narrative provides present and future generations of herpetologists with a basis for a more detailed treatment of the annals of herpetology in the MVZ, perhaps similar to the one recently published for the history of herpetology in the American Museum of Natural History in New York (C. Myers, 2000; see also Resetar and Voris, 1997, for a brief history of herpetology in the Field Museum of Natural History in Chicago). Our account and that of the history of ornithological research in the MVZ (N. Johnson, 1995) briefly chronicle the most salient features of the annals of one of the major vertebrate natural history museums in the United States.

Shortly after the founding of the University of California in 1868 (now the University of California, Berkeley), the brothers John and Joseph Le Conte joined the faculty. John, a geologist, physicist, and naturalist (the long-nosed snake, *Rhinocheilus lecontei*, was named in his honor), became President of the University in November 1868. Joseph, a general natural historian with special interest in glacial geology, became the first Professor of Natural History and in 1869 started lecturing on geology, botany, and zoology (Storer, 1959). Biological studies of amphibians and reptiles at Berkeley began during the Le Conte era. The Department of Zoology was formally established in 1887, and in 1891 William E. Ritter joined Joseph Le Conte and became the first formal Instructor in Zoology. Ritter had broad interests, including the biology of amphibians, and he conducted research on salamanders (Ritter, 1897, 1903), as did his graduate students Loye H. Miller (Ritter and Miller, 1899), Marian E.

1

Hubbard (Hubbard, 1903), and Calvin O. Esterly (Esterly, 1904). Ritter[1] was the founding Editor of the *University of California Publications in Zoology*, which has been a major publication outlet for research by Berkeley herpetologists. Other faculty members in Zoology who sponsored students working on aspects of the biology of amphibians and reptiles in the early twentieth century included Samuel J. Holmes (author of *The Biology of the Frog* [1906], which went through four editions) and J. Frank Daniel (major professor of Richard M. Eakin).[2]

The Museum of Vertebrate Zoology was established on the campus of the University of California in 1908 by the Regents of the University and President Benjamin Ide Wheeler. The force behind the establishment of the MVZ was a remarkable young naturalist, Annie M. Alexander, who conceived the museum and personally selected its first director, Joseph Grinnell (Stein, 2001). By special agreement with the Regents, Alexander personally paid the salaries and expenses of the MVZ and the university provided housing and maintenance of its collections.

Annie Alexander was born in Honolulu, Hawaii, in 1867 and was raised on the family sugarcane farm on Maui. In 1882 her family moved to Oakland, California, where she attended high school. Alexander's father, Samuel T. Alexander, had a strong interest in natural history and was a great traveler. She often accompanied him, and by the time she began planning the museum she had traveled extensively, in Europe, Hong Kong, China, Java, Samoa, the Marquesas Islands, New Zealand, and Africa. She already had begun collecting specimens of various sorts, including fossils, and a chance meeting in 1905 with C. Hart Merriam, renowned mammalogist and head of the Biological Survey in Washington, D.C., stimulated her nascent idea of founding a research museum on the West Coast and encouraged her to travel to Alaska to collect bears. Alexander began planning a museum and continued to collect material for it. In 1907 she met Joseph Grinnell, then teaching at Throop Polytechnic Institute in Pasadena, California (now the California Institute of Technology).[3] Alexander was favorably impressed with

---

[1]In 1903 Ritter established the San Diego Marine Biological Station (now the Scripps Institution of Oceanography), an organization that ultimately led to the University of California, San Diego. In 1909 Ritter left Berkeley and moved to La Jolla to devote full time to the station.

[2]In 1936 Eakin was appointed Assistant Professor at Berkeley's Department of Zoology and served as Chair of the Department from 1942 to 1948. He was an experimental embryologist who conducted research on amphibian development and, later, on the parietal eye of lizards. He wrote two historical accounts of zoological research at Berkeley that covered the periods from 1868 (when the University of California was founded) to 1956 (Eakin, 1956) and from 1957 to 1988 (Eakin, 1988).

[3]In spring 1903 Grinnell's work at Stanford University toward a Ph.D. in zoology was interrupted by a severe attack of typhoid fever. While recovering in Pasadena, he was offered the position of instructor in biology at Throop Polytechnic Institute. He accepted it, with the thought of postponing his return to Stanford for one year, but for unknown reasons, he did not go back to resume his doctoral studies. In 1912 Grinnell wrote to his former major professor at Stanford, Charles H. Gilbert, stating his desire to complete his Ph.D. Gilbert's response was so cordial that Grinnell was encouraged to offer a dissertation. He submitted "An Account of the

Grinnell, especially with his well-organized, private museum located in his home. In 1907 she concentrated her fieldwork in Alaska and assembled a large collection of birds and mammals (H. Grinnell, 1958; Stein, 2001), which became the first specimens accessioned into the MVZ. By fall 1907 plans were well advanced for the Museum of Vertebrate Zoology, and in 1908 Grinnell was appointed Director for a seven-year term that began on April 1.

In 1919, after more than a decade of making monthly contributions for support of the museum, Alexander established a perpetual endowment and made substantial donations to it thereafter. In addition, for the remainder of her life, she continued to contribute money directly for travel, special field trips, equipment, and purchase of specimens. She was consulted on virtually every major activity and field trip, and her insight was greatly responsible for the museum's success. At the time of her death in 1950, Alexander and her longtime field companion, Louise Kellogg,[4] had contributed more than 22,000 specimens of mammals, birds, reptiles, and amphibians to the MVZ collections (Stein, 2001).

The plan Grinnell and Alexander devised for the museum focused on mammals, birds, and reptiles[5] of the West Coast (Alexander, 1907). There were pragmatic as well as idealistic reasons for this decision. Stanford University was the dominant force in ichthyology in the world, and Grinnell had no desire for a competitive engagement with this institution. Further, he believed that one could study the biology of interacting groups of vertebrates at a single time and place by focusing on the assemblage of land vertebrates. In many ways, Grinnell was one of the first community ecologists, but he was also a systematist and a biogeographer. He focused on patterns of local and regional distribution in relation to behavior of the organisms and physical and biotic factors in the environment. He was very interested in geographic variation and systematics, particularly in birds and mammals and mainly at the species and subspecies levels. Thus, he envisioned a museum that concentrated on the assembly of large series of organisms, representing collections through space and time, where variation could be appropriately analyzed and interpreted. During his tenure as Director (his initial seven-year appointment turned into a 31-year stint that ended with his death in 1939), the MVZ adhered stringently to the guidelines laid down in 1908. Grinnell did not lead a single major field expedition outside the borders of California or "Lower California" (i.e., Baja California, which he considered part of a natural region, together with the state of California; Eakin, 1988), even though both he and Alexander had traveled extensively in Alaska and

---

Mammals and Birds of the Lower Colorado Valley with Especial Reference to the Distributional Problems Presented." His thesis was accepted (and published the next year; J. Grinnell, 1914), he took the written examination, and in May 1913 he received his Ph.D. from the President of Stanford University, David Starr Jordan (H. Grinnell, 1940). Simultaneously, Grinnell acquired the title Assistant Professor of Zoology at the University of California (Stein, 2001).

[4]It was considered improper in those days for unmarried women to accompany men in the field without a female companion.

[5]At that time, the term "reptiles" tacitly included amphibians as well.

he maintained a strong interest in that area and its fauna. He was rigidly disciplined with respect to his plan for the museum, which included the following elements: "(1) To ascertain what species of mammals, birds, reptiles and amphibians inhabit western North America, to learn their distributions as accurately as possible, to study their habits, and to attempt to discover the natural laws which control them in various ways. (2) To collect representative individuals of these species and to preserve these as 'specimens' in as permanent a form as is possible, and with full information as to the place, date and circumstances of capture for each specimen. (3) To place on record the results of study of this material. (4) To furnish information, when possible, as to the relation of these forms of native animal life to man's interests" (Storer, 1922).

Grinnell's main interest was birds and secondarily mammals. He was a vigorous researcher, and his productivity was impressive (e.g., J. Grinnell, 1933; J. Grinnell and Camp, 1917; J. Grinnell and Miller, 1944; J. Grinnell and Storer, 1924; J. Grinnell et al., 1930; see H. Grinnell, 1940, for a complete list of J. Grinnell's publications). He also guided several graduate students in their thesis research. His involvement with amphibians and reptiles was substantial early in his career but decreased later. With his wife, Hilda Wood Grinnell, he published on the reptiles of Los Angeles County early in the twentieth century (J. Grinnell and H. Grinnell, 1907).

Grinnell had a great influence on the young Charles L. Camp, whom he met in southern California. Camp was a member of the first museum field trip, to the San Jacinto Mountains of southern California, in 1908. He became an undergraduate student at Berkeley and began working in the collection of amphibians and reptiles in September 1911. Camp left Berkeley in 1915 to begin his doctoral studies at Columbia University in New York with William K. Gregory.[6] In July 1913 another budding herpetologist, Tracy I. Storer, began working in what is still known as the "Herp Lab." Storer, who was a graduate student during the time of his employment (Wake, 1978), is the dominant name in herpetological research in the MVZ during the late teens and early 1920s. His thesis, on the natural history and distribution of the amphibians of California (Storer, 1925), represented a substantial amount of work conducted over many years. Storer became the founding member of the Department of Zoology at what was to become the Davis campus of the University of California. On his departure from the MVZ (July 1, 1923), Margaret Wyeth assumed day-to-day responsibility for the collection of amphibians and reptiles. In addition to Storer, another early museum graduate student who conducted herpetological research was Sarah R. Atsatt (Atsatt, 1913).

During the early period of research on amphibians and reptiles in the MVZ, the collection was housed in a corner of the old museum, occupied in January 1909 and located near what is now known as "faculty glade" on the Berkeley campus. This was a cheaply constructed building that was not

---

[6]Camp returned to Berkeley as an Instructor in Zoology in 1922; in 1930 he became Professor of Paleontology and Director of the Museum of Paleontology on the Berkeley campus and enjoyed a distinguished career as a paleontologist and historian of the American West (Estes, 1988; Wake, 1978).

fireproof and quickly filled with mammal and bird cases. In spring 1930 the new and then modern Life Sciences Building was occupied (Eakin, 1988). A full wing on the west end of this huge building was assigned to the museum. However, the bird and mammal collections dominated the structure, and there was no separate housing of the amphibian and reptile collection until the arrival in 1945 of the first formal Curator of Herpetology, Robert C. Stebbins.

In the 1920s Grinnell, who was promoted to Professor of Zoology in 1920, was the only faculty member associated with the MVZ. The title Curator within the museum organization was in existence rather early, but it acquired formal professorial status only after the mammalogist E. Raymond Hall and the ornithologist Alden H. Miller joined the MVZ staff in 1928 and 1931, respectively. In recent years, curatorial and professorial titles have been strictly parallel, with professorial titles in the Department of Zoology until 1989 and since then in the Department of Integrative Biology.

The policy of the museum since its founding has been to have a common accession system and no formal departmental subdivisions. Accordingly, at various times in the history of the herpetology collection, specialists in mammalogy and ornithology have had supervisory roles in the herpetological collection and have contributed importantly to its development. Although in the early years the museum lacked herpetological faculty and curators, a number of important publications were produced: the first systematic review of the amphibians and reptiles of California (J. Grinnell and Camp, 1917), Camp's descriptions of new taxa (Camp, 1915, 1916b, 1917) and discovery of a Californian species of *Hydromantes* (formerly *Spelerpes*), until then a strictly European salamander genus (Camp, 1916a), and Storer's (1925) *A Synopsis of the Amphibia of California*.

Herpetological activities in the museum were sustained continuously starting in the 1930s, when graduate students began their studies on amphibians and reptiles. Jean M. Linsdale studied the amphibians and reptiles of Baja California and Nevada (Linsdale, 1932, 1940) but was a general vertebrate biologist. He began his long association with the Hastings Natural History Reservation (located in upper Carmel Valley, California, and established in October 1937 by Frances Simes Hastings and Joseph Grinnell) while still a graduate student and spent the rest of his career there (Linsdale, 1943). Henry S. Fitch conducted systematic studies of lizards and snakes as a graduate student (e.g., Fitch, 1934a, b, 1936, 1940) and was influenced permanently by Grinnell's emphasis in the MVZ on studying the natural history of organisms. Fitch enjoyed a long and distinguished career as a herpetological natural historian at the University of Kansas, where he has carried out field studies of reptiles for more than 50 years (Fitch, 1999).[7] A biographical sketch of Fitch (Fitch, 2000) relates some of his experiences as a doctoral student in MVZ in the 1930s. Thomas L. Rodgers, whose association with the museum began in 1935 when he became a graduate student in

---

[7]In 1997 the American Society of Ichthyologists and Herpetologists (publisher of *Copeia*) established an annual award, the Henry S. Fitch Award for Excellence in Herpetology, to recognize an outstanding body of field research by a senior herpetologist.

Berkeley's Department of Zoology, did curatorial work with the collection for several years. Both J. Linsley Gressitt and T. Paul Maslin made collections of amphibians and reptiles in China and published some of the results of their investigations while students in the museum. Gressitt completed his Ph.D. in Entomology and Parasitology and went on to a lengthy and distinguished career as an entomologist in the tropical Pacific region. Maslin, who enjoyed a long career as a herpetologist at the University of Colorado, received his Master's degree at Berkeley but transferred to Stanford University for his doctoral studies.

When Alden H. Miller became the second Director of the Museum in 1940, activity in herpetology increased. This was mainly the result of the appointment of Robert C. Stebbins as Curator of Amphibians and Reptiles and Instructor in Zoology in 1945 and the presence of enthusiastic graduate students such as R. Wade Fox Jr. (e.g., Fox, 1951a, 1954). Stebbins conducted graduate studies at the Los Angeles campus of the University of California, working under the direction of Raymond B. Cowles. He quickly established a major research and graduate education program, and the early part of his career at Berkeley emphasized studies of salamander biology, notably on the ring salamander species *Ensatina eschscholtzii* (Stebbins, 1949, 1954b). Later he studied the physiological ecology of lizards, focusing on the function of the pineal complex (Stebbins, 1960; Stebbins and Cohen, 1973). Perhaps he is best known to the herpetological community and the public for his *A Field Guide to Western Reptiles and Amphibians* (Stebbins, 1966, 1985), the third edition of which he is currently preparing, and for his coauthored book, *A Natural History of Amphibians* (Stebbins and Cohen, 1995). In addition to his research activities, Stebbins worked in biological education at the elementary school level and in conservation biology. He created and maintains the "MVZ Public Service Files," an archival collection of letters, newspaper and magazine articles, and other documents dealing with various conservation issues with which he and other past and present Berkeley faculty members and MVZ graduate students have been involved over several decades. Stebbins has made many important contributions to the MVZ and the academic community at Berkeley. In addition to those mentioned above, these include promoting collecting activities to increase the herpetological holdings of the museum, standardizing the manner in which amphibian and reptilian specimens were preserved, and adopting the use of specimen tags appropriate for long-term storage in ethanol. He was the first faculty member to teach the Herpetology course at Berkeley, and he created the "teaching collection," for which he collected specimens of representative species of various groups, so that students would feel comfortable handling the animals without fear of damaging valuable research material. Nevertheless, he also used museum specimens in his courses. While teaching Herpetology and Vertebrate Natural History, he undertook collecting trips to secure live specimens, which he displayed in naturalistic habitats that he set up in special cages he had designed. On his retirement in 1978, Stebbins was honored with the Berkeley Citation, the highest honor normally given by the University to a faculty member, in recognition of his excellence in teaching, research, and scientific illustration (Eakin, 1988; see *Dedication* for additional biographical information on Stebbins).

The herpetological curatorial staff of the MVZ increased to two with the arrival of David B. Wake. After receiving his Ph.D. under the direction of Jay M. Savage at the University of Southern California in 1964, Wake spent five years as a faculty member at the University of Chicago. He was appointed Associate Curator of Herpetology in the MVZ and Associate Professor of Zoology in 1969, and with two herpetologists the level of activity in herpetology in the MVZ increased. Wake specializes in salamander biology, and on joining the Berkeley faculty he began a long-continued program of field investigations in western North America and Middle America. He is an evolutionary biologist with interests in functional and developmental morphology, molecular systematics, phylogeography, and conservation biology (e.g., Alberch et al., 1979; García-París and Wake, 2000; Lombard and Wake, 1977; Wake, 1991, 1996b, 1997; Wake and Lynch, 1976). Wake was responsible for a dramatic increase in the holdings of amphibians in the MVZ, mainly as a result of his research focus on salamanders from California (USA), México, Guatemala, and Costa Rica. He was noted for offering encouragement and direction to the research programs of the large and dynamic group of graduate students that arrived in the MVZ in the early 1970s and for making the MVZ a preferred destination for several postdoctoral researchers.

Alden Miller died in 1965, and Oliver P. Pearson, a mammalogist who also worked on herpetological topics such as thermoregulation in Andean lizards and toads (Pearson, 1954; Pearson and Bradford, 1976), served as Director of the MVZ until 1971, when David Wake was appointed to the position. Wake was promoted to Professor of Zoology and Curator of Herpetology in 1973. He served as President of several professional societies (Society for the Study of Evolution, American Society of Naturalists, American Society of Zoologists) and was editor of the journal *Evolution* (1978–1980). He was selected Distinguished Herpetologist by the Herpetologists' League in 1984; elected to membership in the American Philosophical Society, the American Academy of Arts and Sciences, and the National Academy of Sciences in 1996, 1997, and 1998, respectively; and received the Joseph Grinnell Medal in Scientific Natural History in 1998 and the Henry S. Fitch Award for Excellence in Herpetology in 1999. He resigned as Director of the Museum in 1998 but remains Curator of Herpetology and Professor of Integrative Biology.

Stebbins officially retired in 1978 (but has remained active), and Harry W. Greene was selected as his replacement. Greene received his Master's degree from the University of Texas, Arlington, in 1973, working under the direction of William F. Pyburn, and his Ph.D. from the University of Tennessee, Knoxville, in 1977, working with Gordon M. Burghardt. He was a faculty member at the University of Pennsylvania for one year before coming to Berkeley. Greene was appointed Assistant Curator of Herpetology and Assistant Professor of Zoology in 1978 and was promoted to Associate Curator of Herpetology and Associate Professor of Zoology in 1983 and to Curator of Herpetology and Professor of Integrative Biology in 1992. His research focused primarily on the natural history of snakes, mainly venomous taxa, a work that he summarized in his acclaimed book, *Snakes: The Evolution of Mystery in Nature* (Greene, 1997), which won the PEN Center West Literary Award for Non-Fiction. He continued the traditions initiated by Stebbins of adding specimens to the collection in the context of field natural history studies and of using MVZ material in undergraduate education and conservation activities. Capitalizing on the MVZ's traditional strength in long series for certain taxa, Greene emphasized the extensive use of museum specimens for studies of lizard and snake feeding ecology, and he used his findings to address broader questions in ecology and evolutionary biology (e.g., Greene, 1986, 1992; Losos and Greene, 1988; Rodríguez-Robles and Greene, 1999). In 1996 he was named Distinguished Herpetologist by the Herpetologists' League. Greene accepted a professorial appointment at Cornell University, New York, and left Berkeley at the end of 1998. In 1999 he was elected President of the American Society of Ichthyologists and Herpetologists and in 2000 received the prestigious Edward Osborne Wilson Naturalist Award of the American Society of Naturalists.

Immediately following Greene's departure, Javier A. Rodríguez-Robles (who received his Ph.D. from Berkeley in 1998 under the direction of Greene) served as Acting Assistant Curator of Herpetology through June, 2000. Rodríguez-Robles studies snake feeding ecology and systematics of western North American and West Indian reptiles. During his curatorial tenure, he focused on updating the taxonomy and classification of the collection and on making emendations to the MVZ's online herpetological catalog.

Craig Moritz, from the University of Queensland, Australia, was appointed Professor of Integrative Biology, Curator of Herpetology, and Director of the Museum of Vertebrate Zoology and began his service in January 2001. Moritz worked extensively with Australian geckos early in his career but has expanded his interest to many other taxa in recent years. He is an evolutionary biologist and an active participant in population genetics, phylogeography, and conservation biology. Between the period of Wake's resignation (September 1998) and Moritz's arrival, James L. Patton, Curator of Mammals and Professor of Integrative Biology, served as Acting Director of the museum.

Several Berkeley faculty members with active research programs in herpetology have been associated with the MVZ in various capacities. Marvalee H. Wake received her Ph.D. from the University of Southern California in 1968, working with Jay M. Savage. She joined the Berkeley faculty in 1969 and was appointed Research Morphologist in the MVZ in 1975. She subsequently became Chair of the Department of Zoology and also served on two occasions as the Chair of its successor, the Department of Integrative Biology. An evolutionary, comparative, and developmental morphologist and systematist, she specializes in caecilian biology and has a strong interest in evolutionary reproductive biology. She has sponsored various MVZ herpetological graduate students and in 1969 cofounded with David Wake an ongoing, popular weekly to biweekly (in recent years) seminar series on the biology of amphibians and reptiles informally known as "Herp Group." In 1983 she served as President of the American Society of Ichthyologists and Herpetologists and was the Secretary-General of the Third World Congress of

Herpetology held in Prague, Czech Republic, in 1997. She is currently the President of the Society of Integrative and Comparative Biology (until January 2003) and was elected President of the International Union of Biological Sciences in 2000. Paul Licht, a comparative endocrinologist who works on amphibians and reptiles (Ph.D., University of Michigan, 1964), worked with faculty and students of the MVZ and sponsored herpetological graduate students. He served as Chair of the Department of Zoology and currently is Dean of Biological Sciences and Chair of Deans in the College of Letters and Science. Tyrone B. Hayes (who received his Ph.D. from Berkeley in 1993, under the direction of Licht) is a developmental endocrinologist and an Associate Professor of Integrative Biology. He is also Associate Research Developmental Biologist in the MVZ and has sponsored herpetological graduate students.

Curatorial Associate Barbara R. Stein (who received her Ph.D. in Systematics and Ecology from the University of Kansas in 1985) directed daily operations in the herpetological laboratory from 1985 until June 2000, when she relocated to Seattle, Washington. She played a similar role in the mammal collection and also was responsible for electronic database management for the museum as a whole. Curatorial Associate Carla Cicero (Ph.D. in Wildland Resource Science from Berkeley in 1993) is in charge of the tissue collection. In July 2001 Christopher J. Conroy (Ph.D. in Biology from the University of Alaska in 1998) succeeded Stein and currently oversees the mammal and herpetological collections.

Several postdoctoral Curatorial Associates were appointed for varying periods of time, mainly in conjunction with sabbatical leaves and with a National Science Foundation grant for computerization of the MVZ herpetological collections, which was completed in 1989. These included Allen E. Greer (who received his Ph.D. under the direction of Ernest E. Williams at Harvard University and spent a year at Berkeley before joining the Australian Museum in Sydney as a Curator), Richard D. Sage (a Berkeley undergraduate who received his Ph.D. under Frank Blair at the University of Texas, Austin, and who served as supervisor of the museum's evolutionary genetics laboratory and as an Associate Researcher before taking a professorial position at the University of Missouri, Columbia), and David A. Good (who received his Ph.D. from Berkeley in 1985, under the direction of David Wake, and who moved to a postdoctoral position in the Redpath Museum at McGill University in Montreal, Canada, after leaving the MVZ). Good was responsible for the four-year program of computerization, which included an exhaustive and comprehensive review of the entire collection, during which the identification and data of all specimens were checked. Stephen D. Busack (Ph.D. from Berkeley in 1985 under the direction of David Wake) was a long-term Curatorial Assistant in day-to-day charge of the herpetological collections from 1978 to 1984. Sage, Good, and Busack were active research scientists during their time in the MVZ, with respect to both fieldwork and morphological and molecular systematic studies.

Since the 1970s the level of herpetological activity in the MVZ has been high, with many doctoral and postdoctoral students (see *Appendix*). Several graduate students who conducted their dissertation research while in residence in the MVZ and who obtained their doctoral degrees from Berkeley hold or have held professorial appointments at prestigious institutions such as Louisiana State University, Baton Rouge; Museum of Comparative Zoology at Harvard University; Pontificia Universidad Católica de Chile, Santiago; University of Kansas, Lawrence; University of Colorado, Boulder; University of Connecticut, Storrs; University of Texas, Austin; and Washington University, St. Louis. Other former MVZ students hold or have held curatorial appointments at the Academy of Natural Sciences of Philadelphia, the Field Museum of Natural History in Chicago, and the National Museum of Natural History in Washington, D.C.

Accession number six (May–October 1908) of the Museum of Vertebrate Zoology, from the vicinity of the San Jacinto Mountains in southern California, was the first to contain amphibians and reptiles. The first specimen entered on the herpetological catalog (MVZ 1), on March 13, 1909, was a *Crotaphytus bicinctores* from Cabazon, Riverside County, California, collected by Walter P. Taylor on May 9, 1908. Early entries in the amphibian and reptile catalog were by Joseph Grinnell and a longtime assistant, Harry S. Swarth. When Robert Stebbins joined the museum in 1945, there were about 41,000 specimens in the collection; when David Wake arrived in 1969, there were nearly 85,000; at the time of Harry Greene's appointment in 1978, the herpetological collection had grown to more than 151,000 specimens; and when Craig Moritz joined the MVZ in 2001, the collection had almost 231,000 specimens. Collecting activities in the museum reached a peak during the 1970s and early 1980s (Fig. 2), when many of the most active contributors were working in Middle and South America. Among those who have contributed large collections to the museum (in addition to the Curators) are John E. Cadle, David A. Good, James F. Lynch, J. Robert Macey, Theodore J. Papenfuss, Richard D. Sage, Robert L. Seib, Samuel S. Sweet, and Kay P. Yanev. Theodore J. Papenfuss (who received his Ph.D. from Berkeley under the direction of Robert Stebbins in 1979) has been a Research Associate in the MVZ since 1979 and is associated mainly with the museum's field program, most recently in China, Russia, México, and central Asian and eastern African countries. He has published extensively in the field of molecular systematics. Anita K. Pearson, whose research interests include embryology and endocrinology, has also been associated with the museum since 1979.

In the early 1970s the museum organized a laboratory of evolutionary genetics, and the first supervisor was Suh Yung Yang, who received his Ph.D. with Robert K. Selander at the University of Texas, Austin. A frozen tissue collection was organized that has since grown to more than 16,500 herpetological specimens. Yang returned to his native South Korea for a professorial appointment, where he has remained active in herpetology. The evolutionary genetics laboratory began sequencing DNA in the late 1980s and has become an integral part of the museum's research program.

The biological sciences program at Berkeley began a major reorganization in the mid-1980s, accompanied by a building renovation and modernization plan. The Department of Zoology was discontinued in 1989, and since then the home department of MVZ faculty and students has been its successor, the Department of Integrative Biology. The Life Sciences Building was vacated in

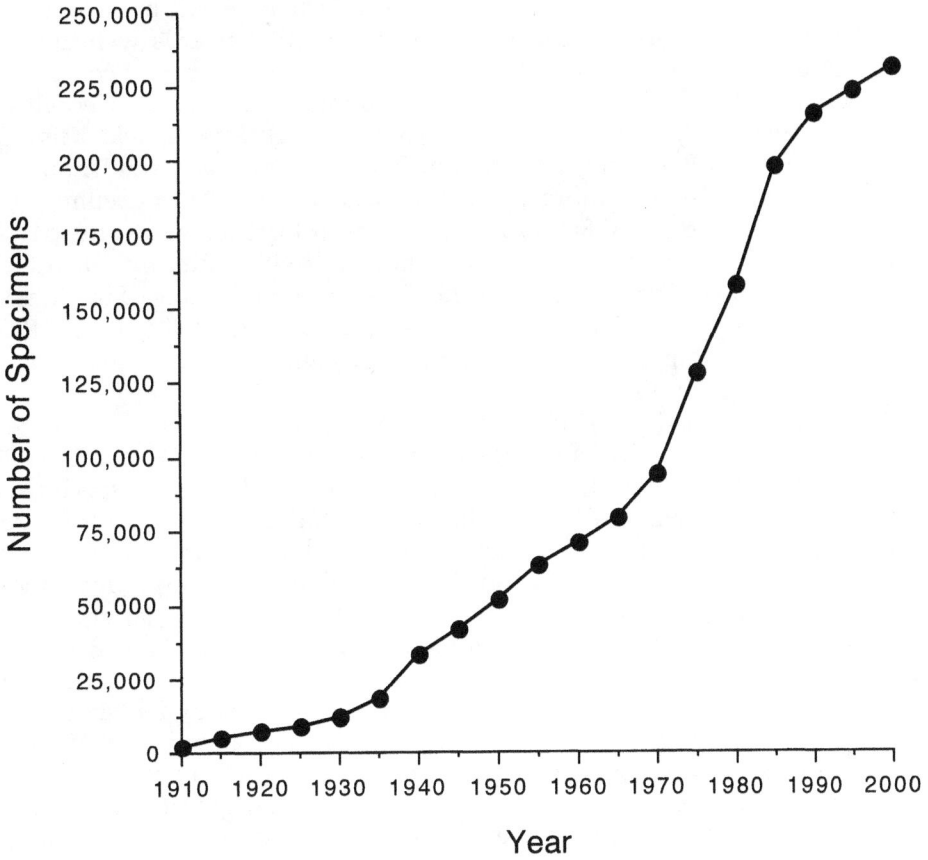

Figure 2. Growth of the herpetological collection in the Museum of Vertebrate Zoology.

1990 and a major renovation was begun. For more than three years the MVZ was forced into a portion of its previous space as construction proceeded. In fall 1993 the museum collections were moved into new and expanded quarters on the third floor of the renamed Valley Life Sciences Building, and in January 1994 the museum was fully functional once again.

Starting in the late 1960s, the MVZ became a premier destination for postdoctoral students in herpetology. In large part this resulted from the establishment in 1957 of the Miller Institute for Basic Research in Science, an independent organization on the Berkeley campus that awards well-funded two- to three-year postdoctoral fellowships. The following have been Miller Fellows in the MVZ: Stevan J. Arnold, Edmund D. Brodie III, George C. Gorman, Raymond B. Huey, Robert H. Kaplan, Mark Kirkpatrick, Kiisa C. Nishikawa, Steven Poe, Michael J. Ryan, Paul W. Sherman, and Adam P. Summers. Miller Fellows more loosely associated with the museum were Albert F. Bennett and Barry Sinervo. Other postdoctoral fellows in the MVZ in herpetology have been João M. Alexandrino, Kellar Autumn, David C. Cannatella, David Carrier, M. Brent Charland, Jinzhong Fu, Mario García-París, David M. Green, Hugh Griffith, Enrique P. Lessa, R. Eric Lombard, Matilde Ragghianti, Leslie J. Rissler, Javier A. Rodríguez-Robles, and Kelly R. Zamudio.

The herpetological collection in the MVZ is currently under the supervision of Curators David B. Wake and Craig Moritz. More than 232,000 specimens (as of December 31, 2001), which include over 1,700 type specimens, are accessioned and cataloged. The collection (including approximately 4,000 cleared and stained skeletal preparations) is housed in six specially constructed fluid rooms, and the sizable osteological holdings (over 5,300 specimens) are housed in a separate room. The herpetological collection differs from that of many other museums in that it was not planned to be a synoptic collection. Rather, it is a manifestation of the research activity of generations of associated students and academic and nonacademic staff. The MVZ was founded with the philosophy that organisms should be studied in relationship to their environments, and therefore the specimen collections are supplemented by extensive field notes, maps, photographs, and correspondence, which enhances their value to researchers. In fact, although the many large series in the MVZ have been used extensively by systematists, they have also served for many kinds of ecological, biogeographic, and life history studies. Furthermore, in recent years numerous phylogeographic studies have been conducted using museum materials and facilities, and we expect this trend to continue as the burgeoning new field of biodiversity science matures. This multiplicity of uses, not only of the herpetological collection, but also of the ornithological and mammal holdings in the MVZ, was foreseen almost a century ago by the museum's first director, Joseph Grinnell, who in 1910 wrote, "It will be observed, then, that our efforts are not merely to accumulate as great a mass of animal remains as possible. On the contrary, we are expending even more time than would be required for the collection of the specimens alone, in rendering what we do obtain as permanently valuable as we know how, to the ecologist as well as the systematist. It is quite probable that the facts of distribution, life history, and economic status may finally prove to be of more far-reaching value, than whatever information is obtainable exclusively from the specimens themselves" (J. Grinnell, 1910:165–166). Thus, the MVZ, in addition to embracing new research approaches and technologies, continues to adhere to its long-established philosophy.

Service to our biological colleagues and the general public will continue to be a main theme in the MVZ, as witnessed by the fact that we were one of the first museums to make our complete catalog available on the World Wide Web (http://elib.cs.berkeley.edu/mvz/). In years to come, we expect the herpetological program in the MVZ to continue to be innovative and productive.[8]

---

[8]A list (exclusive of abstracts and book reviews) of the more than 1,300 (as of December 2001) articles, book chapters, and books on various aspects of the biology of amphibians and nonavian reptiles published by researchers associated with the MVZ since its founding in 1908 is available online at http://www.mip.berkeley.edu/mvz/collections/MVZHerpPubs.htm.

TYPE SPECIMENS OF RECENT
AMPHIBIANS AND NONAVIAN REPTILES
IN THE MUSEUM OF VERTEBRATE ZOOLOGY

The herpetological collection of the Museum of Vertebrate Zoology, University of California, Berkeley, was begun on March 13, 1909, with a collection of approximately 430 specimens from southern California and as of December 31, 2001, contained 232,254 specimens. Taxonomically, the collection is strongest in salamanders, accounting for 99,176 specimens, followed by "lizards" (squamate reptiles other than snakes and amphisbaenians, 63,439), frogs (40,563), snakes (24,937), turtles (2,643), caecilians (979), amphisbaenians (451), crocodilians (63), and tuataras (3). Whereas the collection's emphasis historically has been on the western United States and on California in particular, representatives of taxa from many other parts of the world are present. Eleven countries are represented by more than 1,000 specimens each (Argentina, China, Colombia, Costa Rica, Guatemala, Jamaica, México, Perú, South Africa, the United States, and Vietnam). The United States material accounts for more than 60% of the MVZ collection, with 142,181 specimens, followed by México with 27,870 and Guatemala with 18,583. Within the United States, the regional nature of the MVZ collection is apparent in the distribution of specimens among the states, the top five being California (90,457), Oregon (7,823), Nevada (6,782), Texas (6,084), North Carolina (5,206), and Arizona (5,185), with California accounting for approximately 64% of the total specimens from the country.

Every institution in which name-bearing specimens are deposited should publish a catalog of the type specimens in its care (Recommendation 72F.4, International Commission on Zoological Nomenclature, 1999). In 1962 Robert G. Crippen published a list of the 54 holotype specimens then present in the MVZ herpetological collection (Crippen, 1962). In the intervening

39 years, 66 additional MVZ holotypes were designated or discovered, and it is therefore appropriate to update Crippen's list. In addition, Crippen limited his discussion to holotypes and made no mention of other type material. Herein we include all type material (holotypes, neotypes, syntypes, paratopotypes, and paratypes) known to be present in the MVZ herpetological collection. Of the 196 amphibian and nonavian reptilian taxa represented by type specimens in the MVZ, most were collected in México (63) and California (USA, 54). These are followed by Guatemala (15), Perú (9), Costa Rica and Honduras (each with 6), China and Oregon (USA) (each with 5), Arizona (USA), Taiwan, and Vietnam (each with 4), Colombia, New Mexico (USA), Texas (USA), and Virginia (USA) (each with 2), and Ecuador, El Salvador, Idaho (USA), Iran, the Marshall Islands, Morocco, North Carolina (USA), Panamá, Papua New Guinea, the Solomon Islands, Tanzania, Utah (USA), and Washington (USA) (each with 1).

Tables 1 and 2 summarize the known type material in the MVZ. For many years paratype material was not accorded special treatment; it was stored on the shelves with the ordinary specimens, and no accurate records of the presence of such material were kept. The following list is therefore probably incomplete in that paratype material that is unknown to us may exist in the collection. The MVZ would greatly appreciate being informed of any type specimens in our collection that are not listed herein.

In this list the species accounts are arranged alphabetically by clade (i.e., Anura, Caudata, Gymnophiona, Squamata, Serpentes, Testudines), family, and species, in that order. When the author(s) of a scientific name is (are) not the same as that (those) of the type citation, or when it was necessary to distinguish between authors with identical last names, the appropriate names appear in brackets before the type citation. All type localities, the geographic place of capture of holotypes, neotypes, syntypes, and paratopotypes (paratype specimens collected at the type locality), are given in quotation marks as they appear in the original description. Any clarifications, corrections, or incongruences regarding the type locality resulting from our examination of specimen tags, field notes, or the relevant scientific literature are set off in brackets or indicated immediately after the published locality data. For articles that appeared the year after they are dated, the year in which they became available is followed by the year (in parentheses) indicated in the article. When it was necessary to add a letter after the publication year of an article to distinguish between papers published by the same author(s) in the same year, the letter appears in brackets to indicate that it is not part of the type citation of the name in question. When the collector assigned a field number to a specimen that was later designated a holotype or a neotype, this number appears in parentheses following the collector's name. The locality data for all paratypes are as specific as possible, include clarifications and corrections to the localities published in the original description, and follow a standardized format. Any incongruent or relevant information about the type specimens is noted in the Remarks section. Museum acronyms used are AMNH = American Museum of Natural History, New York; CIB = Chengdu Institute of Biology, Chengdu, Sichuan, China; CPS = University of Puget Sound, Tacoma, Washington; LACM = Los Angeles County Museum of

Natural History, California; LSUMZ = Louisiana State University, Museum of Zoology, Baton Rouge; MCZ = Museum of Comparative Zoology, Harvard University, Cambridge, Massachusetts; SU = Stanford University Natural History Museum, Palo Alto, California (this collection now resides in the California Academy of Sciences, San Francisco [CAS-SU]); UCLA = Department of Biology, University of California, Los Angeles (this collection now resides at LACM); UCR = Museo de Zoología, Universidad de Costa Rica, San José; UMMZ = University of Michigan Museum of Zoology, Ann Arbor; UNAH = Universidad Nacional Autónoma de Honduras, Tegucigalpa; USNM = National Museum of Natural History, Smithsonian Institution, Washington, D.C.; and UTA = Collection of Vertebrates, University of Texas, Arlington. Other abbreviations used are ca. = approximately, Co. = County, ft = feet, and mi = mile.

TABLE 1. Summary of herpetological type material in the Museum of Vertebrate Zoology (as of December 31, 2001). "Lizards" refers to squamate reptiles other than amphisbaenians and snakes. Numbers in parentheses indicate missing specimens.

	Holotypes	Neotypes	Syntypes	Paratopotypes and Paratypes
Frogs	11	0	0	77 (1)
Salamanders	62	1	3	1,334 (4)
Caecilians	0	0	0	1
"Lizards"	20	2	0	172
Amphisbaenians	1	0	0	11
Snakes	25	0	0	35 (1)
Turtles	1	0	0	9
Totals	120	3	3	1,639 (6)

TABLE 2. Summary of taxa of herpetological holotypes, neotypes, and syntypes (as described originally) in the Museum of Vertebrate Zoology (as of December 31, 2001). "Lizards" refers to squamate reptiles other than amphisbaenians and snakes.

	Species	Subspecies
Frogs	6	5
Salamanders	58	6*
"Lizards"	10	12
Amphisbaenians	1	0
Snakes	8	17
Turtles	0	1
Totals	83	41

*Because one of these taxa is represented by two additional syntypes, the row total, 64, does not match the total number of holotypes, neotypes, and syntypes of salamanders in Table 1.

LIST OF TYPE SPECIMENS

AMPHIBIA

ANURA

Ascaphidae

Ascaphus truei californicus Mittleman and Myers

Proc. Biol. Soc. Washington 62:63. 1949.
CURRENT NAME: *Ascaphus truei* Stejneger
HOLOTYPE: **MVZ 19142** (female); "near Klamath, Del Norte County, Calif. [=California,] [USA]"; W. F. Wood (33-1692); November 4, 1933.
PARATYPES: **MVZ 29790-29793, 29795, 29797-29798, 29801-29803**; tributary of Wilson Creek, 8.5 mi N of Klamath, Del Norte County, California, USA.
REMARKS: MVZ 29801 is a tadpole.

Ascaphus truei montanus Mittleman and Myers

Proc. Biol. Soc. Washington 62:64. 1949.
CURRENT NAME: *Ascaphus montanus* Mittleman and Myers
PARATYPES: **MVZ 12336, 12344**; 1 mi NE of Heath, on southwestern slope of Cuddy Mountain, Washington County, Idaho, USA, 4,000 ft elevation. **MVZ 12340-12343, 12345**; 0.5 mi E of Black Lake, Adams County, Idaho, USA, 6,000 ft elevation.

Bufonidae

Bufo canorus Camp

Univ. California Publ. Zool. 17:59. 1916[b].
HOLOTYPE: **MVZ 5744** (adult female); "Porcupine Flat, 8100 feet altitude, Yosemite National Park, Mariposa County, California[, USA]"; C. L. Camp (2129); July 1, 1915.

Bufo cognatus californicus Camp

Univ. California Publ. Zool. 12:331. 1915.
CURRENT NAME: *Bufo californicus* Camp
HOLOTYPE: **MVZ 4364** (adult female); "Santa Paula, 800 feet altitude, Ventura County, California[, USA]"; C. L. Camp (551); May 22, 1912.

Bufo exsul Myers

[G. Myers.] Occ. Pap. Mus. Zool., Univ. Michigan 460:3. 1942[b].
PARATOPOTYPES: **MVZ 26048-26049**; "Deep Springs, Deep Springs Valley, Inyo County, California, [USA]."

Bufo spiculatus Mendelson

Herpetologica 53:269. 1997[b].
PARATYPE: **MVZ 146824**; México Highway 175, 7.7 mi (12.4 km) S of La Esperanza, Sierra de Juárez, Oaxaca, México, 1,980 m elevation.
REMARKS: MVZ 14682 (*Plethodon cinereus*) was erroneously listed as a paratype in the original description. MVZ 146824 matches the data presented by Mendelson and therefore is believed to be a paratype. In addition, loan records in the MVZ show that Mendelson borrowed the latter specimen, not MVZ 14682.

Bufo tutelarius Mendelson

Herpetologica 53:15. 1997[a].
PARATYPES: **MVZ 113616**; 1.5 km (by air) SE of San Rafael Pie de la Cuesta, Finca Santa Julia, Departamento San Marcos, Guatemala, 1,075 m elevation. **MVZ 165511, 180391**; 1.25 km E, 0.75 km S of San Rafael Pie de la Cuesta, Finca Santa Julia, Departamento San Marcos, Guatemala, 1,100 m elevation.

Hylidae

Hyla californiae Gorman

Herpetologica 16:214. 1960.
CURRENT NAME: *Hyla cadaverina* Cope; *Pseudacris cadaverina* (Cope) of some authors
HOLOTYPE: **MVZ 31773** (adult female); "Canyon de Llanos, 9 mi. (14.5 km.) SSW of 'Alaska' (La Rumorosa), Partido del Norte, Baja California, Mexico"; R. R. Miller and J. Davis; June 10, 1939.

Hyla minera Wilson, McCranie, and Williams

Herpetologica 41:145. 1985.
HOLOTYPE: **MVZ 130790** (adult male); "4.2 km (by road) S [of] Purulhá, 1760 m elevation, Depto. [=Departamento] Baja Verapaz, Guatemala"; J. F. Lynch (G-9740), T. J. Papenfuss, and D. B. Wake; November 14, 1974.

Leptodactylidae

Ceratophrys stolzmanni scaphiopeza Peters

Proc. Biol. Soc. Washington 80:105. 1967.
PARATYPE: **MVZ 77182**; Playas, 60 mi SW of Guayaquil, Provincia Guayas, Ecuador.
REMARKS: MVZ 77182 was erroneously referred to as "UCMVZ 77182" in the original description.

Eleutherodactylus catalinae Campbell and Savage

Herpetol. Monogr. 14:276. 2000.
PARATYPE: **MVZ 219598**; woods beside Río Cotón, below "La Casita," ca. 2 km ESE of Las Tablas, Provincia Puntarenas, Costa Rica, 1,880-1,910 m elevation.

Eleutherodactylus daryi Ford and Savage

Occ. Pap. Nat. Hist. Mus., Univ. Kansas 110:1. 1984.
PARATYPES: **MVZ 104648-104649**; cloud forest above Finca Chichén, 12 km (by air) due S of Cobán, Departamento Alta Verapaz, Guatemala, ca. 1,850 m elevation. **MVZ 130784-130786**; cloud forest along CA-14, 4.2 km (by road) S of Purulhá, Departamento Baja Verapaz, Guatemala, 1,760 m

elevation. **MVZ 144529**; 4 km ENE of Chilascó, Departamento Baja Verapaz, Guatemala, 1,829 m elevation. **MVZ 150933**; Finca San Jorge, 4 km NE of Chilascó, Departamento Baja Verapaz, Guatemala, 1,900 m elevation. **MVZ 150936**; 5 km NE of Chilascó, Departamento Baja Verapaz, Guatemala, 1,900 m elevation. **MVZ 175813**; edge of cloud forest along Ruta Nacional 5, 2.5-4.8 km S of Purulhá, Departamento Baja Verapaz, Guatemala, 1,630-1,720 m elevation.

REMARKS: MVZ 175813 is missing.

Eleutherodactylus omoaensis McCranie and Wilson

Alytes 14:155. 1997.

HOLOTYPE: **MVZ 115286** (adult male); "about 10 airline km WSW [of] San Pedro Sula on road to Perú (15° 28′ N, 88° 06′ W), elevation 1150 m, Sierra de Omoa, Departamento de Cortés, Honduras"; J. Kezer and J. F. Lynch; February 9, 1974.

PARATOPOTYPES: **MVZ 115281-115285, 115287-115288, 115290, 128749-128752.**

Megophryidae

Leptolalax sungi Lathrop, Murphy, Orlov, and Ho

Amphibia-Reptilia 19:254. 1998.

PARATOPOTYPE: **MVZ 223699**; "a stream on the east side of the village of Tam Dao, (21° 27′ 31″ N; 105° 38′ 61″ E), elevation 925 m, Vinh Phu Province, Vietnam."

REMARKS: The species name was erroneously spelled *"Lepolalax sungi"* in one part of the original description.

Microhylidae

Kaloula pulchra hainana Gressitt

Proc. Biol. Soc. Washington 51:127. 1938[a].

HOLOTYPE: **MVZ 23189** (adult female); "Kachek [=Qionghai County], alt. [=altitude] 25 meters, eastcentral Hainan Island [=Hainan Province], South China Sea[, China] (lat. [=latitude] 18° 50′ N., long. [=longitude] 110° 30′ E.)"; J. L. Gressitt; August 7, 1935.

PARATOPOTYPE: **MVZ 23188.**

Otophryne pyburni Campbell and Clarke

Herpetologica 54:309. 1998.
PARATYPES: **MVZ 200479-200480**; Yapima, Departamento Vaupés, Colombia, 145 m elevation (ca. 01° 01' N, 69° 27' W).
REMARKS: MVZ 200479-200480 were formerly UTA A-4381 and UTA A-12789, respectively.

Rana gracilipes Gressitt

Proc. Biol. Soc. Washington 51:161. 1938[b].
CURRENT NAME: *Micryletta steinegeri* (Boulenger)
HOLOTYPE: **MVZ 23108** (adult female); "Kuraru [=Kenting Forestry Park], alt. [=altitude] 150 meters, Koshun district [=Hengchun township], near South Cape, [Pingtung County,] Formosa [=Taiwan, "Republic of China"]"; J. L. Gressitt (197); August 10, 1934.

Ranidae

Rana attigua Inger, Orlov, and Darevsky

Fieldiana, Zool. (New Ser.) 92:14. 1999.
PARATOPOTYPES: **MVZ 221932-221934, 221936, 221941-221943**; "Buon Luoi, An Khe District, Gia Lai Province, Vietnam."
REMARKS: MVZ 222932-222934 (*Batrachoseps major*), 222936 (*B. major*), and 222941-222943 (*Batrachoseps* sp. nov.) were erroneously listed as paratypes in the original description. MVZ 221932-221934, 221936, and 221941-221943 match the data presented by Inger et al. and therefore are believed to be paratopotypes. In addition, loan records in the MVZ show that Inger borrowed the latter specimens, not MVZ 222932-222934, 222936, and 222941-222943.

Rana boylii muscosa Camp

Univ. California Publ. Zool. 17:118. 1917.
CURRENT NAME: *Rana muscosa* Camp
HOLOTYPE: **MVZ 771** (adult female); "Arroyo Seco Cañon, at about 1300 feet altitude, near Pasadena, [Los Angeles County,] California[, USA]"; J. Grinnell; August 3, 1903.

Rana boylii sierrae Camp

Univ. California Publ. Zool. 17:120. 1917.
CURRENT NAME: *Rana muscosa* Camp
HOLOTYPE: **MVZ 3734** (adult female); "Matlack Lake, 10,500 feet altitude, two miles southeast of Kearsarge Pass, Sierra Nevada, Inyo County, California[, USA]"; H. S. Swarth (9901); June 26, 1912.

Rana pueblae Zweifel

Univ. California Publ. Zool. 54:253. 1955.
PARATOPOTYPES: **MVZ 60240-60241**;   "2.8   miles   northeast   of Huauchinango, Río Texcapa, Puebla, México."
REMARKS: MVZ 60240 and 60241 were formerly UMMZ 99473 and 99469, respectively. The original description lists the acronym of these specimens as "UMNZ," but this is in error.

Rana sinaloae Zweifel

Bull. Southern California Acad. Sci. 53:131. 1954[a].
CURRENT NAME: *Rana pustulosa* Boulenger
HOLOTYPE: **MVZ 58962** (adult female); "14 miles by road southwest of El Batel, Sinaloa, Mexico, at an elevation of 4200 feet.   El Batel is about 33 miles air line distance east and 15 miles north of Mazatlán"; R. G. Zweifel (2478); July 12, 1953.
PARATOPOTYPES: **MVZ 58959-58961, 58963-58965**.
PARATYPE: **MVZ 58966**; 10 mi (by road) NE of El Batel, Sinaloa, México, 6,400 ft elevation.

Rana tlaloci Hillis and Frost

Occ. Pap. Nat. Hist. Mus., Univ. Kansas 117:10. 1985.
PARATOPOTYPE: **MVZ 186221**; "Xochimilco, Distrito Federal, México." Field notes of the collector, H. Bradley Shaffer, indicate that MVZ 186221 (field number HBS 4709) was purchased at a marketplace in Xochimilco, but it is not known whether the vendor collected the specimen locally.

Rhacophoridae

Philautus abditus Inger, Orlov, and Darevsky

Fieldiana, Zool. (New Ser.) 92:26. 1999.
PARATOPOTYPES: **MVZ 222101, 222118-222121**; "Buon Luoi, An Khe District, [Gia Lai Province,] Vietnam."

Rhacophorus baliogaster Inger, Orlov, and Darevsky

Fieldiana, Zool. (New Ser.) 92:30. 1999.
PARATOPOTYPES: **MVZ 222040-222042, 222100, 222104**; "Buon Luoi, An Khe District, [Gia Lai Province,] Vietnam."

CAUDATA

Ambystomatidae

Ambystoma macrodactylum croceum Russell and Anderson

Herpetologica 12:137. 1956.
HOLOTYPE: **MVZ 63734** (adult male); "Rio Del Mar, Santa Cruz County, California[, USA]"; J. D. Anderson (1106) and R. W. Russell; December 2, 1954.
PARATOPOTYPES: **MVZ 63726-63733, 63735-63736**.
REMARKS: The original description referred to several paratypes but did not list them explicitly. MVZ 63726-63733 and 63735-63736 match the data for those discussed by Russell and Anderson and therefore are believed to be paratopotypes.

Ambystoma tigrinum utahense Lowe

Trans. Kansas Acad. Sci. 58:246. 1955.
CURRENT NAME: *Ambystoma tigrinum nebulosum* Hallowell
HOLOTYPE: **MVZ 29481** (adult male); "Lapoint, Uintah County, Utah[, USA]"; D. L. Bills (W. F. Wood 35-124); May 1935.

Hynobiidae

Batrachuperus taibaiensis Song, Zeng, Wu, Liu, and Fu

Asiatic Herpetol. Res. 9:6. 2001.
PARATOPOTYPES: **MVZ 230964-230965, 230979-230986**; "upper stream of Heihe River, near Hua Er Ping Village, Zhouzhi County, Shaanxi Province, China (33.85° N, 107.82° E), elevation 1260 m."

Plethodontidae

Aneides flavipunctatus niger Myers and Maslin

Proc. Biol. Soc. Washington 61:132. 1948.
PARATYPES: **MVZ 8261**; Los Gatos, Santa Clara County, California, USA. **MVZ 12207-12208**; 20 mi SW of San Jose, Santa Clara County, California, USA. **MVZ 13801-13806, 16052**; Ben Lomond, Santa Cruz County, California, USA. **MVZ 18002-18003**; head of Mindego Creek, San Mateo County, California, USA. **MVZ 27805-27853, 27858**; 1.5 mi W of Felton, Santa Cruz County, California, USA, 1,000 ft elevation.
REMARKS: MVZ 16053 (*Diadophis punctatus*), 27854-27856 (*Bufo boreas*), and 27857 (*Sceloporus occidentalis*) were erroneously listed as paratypes in the original description. MVZ 27810, 27812, 27842, and 27858 are cleared and stained skeletal preparations.

Aneides vagrans Wake and Jackman

[Wake and Jackman, in Jackman.] Can. J. Zool. 76:1579. 1999 (1998).
HOLOTYPE: **MVZ 124876** (adult male); "about 10 km S [of] Maple Creek, Humboldt Co., California, [USA,] 40° 42′ N, 123° 50′ W, ca. 500 m elevation"; J. F. Lynch, L. D. Houck, and M. Stevens; March 8, 1970.
PARATOPOTYPES: **MVZ 124877-124883**.

Batrachoseps campi Marlow, Brode, and Wake

Contrib. Sci., Nat. Hist. Mus. Los Angeles County 308:3. 1979.
HOLOTYPE: **MVZ 122993** (adult female); "Long John Canyon, W slope of the Inyo Mountains, elevation 1695 m (5560 ft), 3.2 km (2 mi) (airline) N, 5.3 km (3.3 mi) E [of] Lone Pine, Inyo County, California, USA"; R. W. Marlow (B-001) and J. M. Brode; September 26, 1973.
PARATOPOTYPES: **MVZ 122994-122996, 123011-123015**.

PARATYPES: **MVZ 122997-123009, 123017-123031**; French Spring, western slope of Inyo Mountains, 4 mi (6.4 km) N, 3.6 mi (5.7 km) E of Lone Pine, Inyo County, California, USA, 6,000 ft (1,829 m) elevation.

REMARKS: MVZ 123031 is a cleared and stained skeletal preparation.

Batrachoseps diabolicus Jockusch, Wake, and Yanev

Contrib. Sci., Nat. Hist. Mus. Los Angeles County 472:7. 1998.

HOLOTYPE: **MVZ 95446** (adult male); "Hell Hollow, at the junction with the Merced River at Lake McClure on California Highway 49, Mariposa County, California, [USA,] 37° 36′ 33″ N, 120° 08′ 10″ W, ca 300 m elevation"; S. B. Ruth (1078); February 6, 1971.

PARATOPOTYPES: **MVZ 95444-95445, 95449-95451, 95454-95455, 156572, 156586, 156588, 156598-156599, 156601, 156603-156604, 224836-224838.**

Batrachoseps gabrieli Wake

Contrib. Sci., Nat. Hist. Mus. Los Angeles County 463:1. 1996[a].

HOLOTYPE: **MVZ 196449** (adult female); "slope above Soldier Creek in the upper San Gabriel River drainage, approximately 1 km ESE [of] Crystal Lake, San Gabriel Mountains, Los Angeles County, California[, USA]. SW Section 28, R9W T3N. 34° 18′ 47″ N, 117° 49′ 57″ W[, a]pproximately 1,550 m elevation"; D. B. Wake (1790), N. Staub, S. S. Sweet, A. Tate, S. G. Tilley, and J. Tilley; March 28, 1985.

PARATOPOTYPES: **MVZ 178631-178646, 195577-195583, 196450-196463, 215938, 215940-215946, 215948, 222957-222961.**

REMARKS: MVZ 178632, 178634, 178639, 178642, and 195582-195583 are cleared and stained skeletal preparations.

Batrachoseps gavilanensis Jockusch, Yanev, and Wake

Herpetol. Monogr. 15:69. 2001.

HOLOTYPE: **MVZ 155642** (adult female); "0.5 miles (0.8 km) south of cement plant on San Juan Creek Road, San Benito Co., CA [=California], [USA,] 36° 49′ 13″ N, 121° 31′ 30″ W, [e]levation ca. 120 m"; K. P. Yanev and S. S. Sweet; January 21, 1973.

PARATOPOTYPES: **MVZ 155637-155641, 155643, 155645, 155690, 155699-155702, 155704-155706, 155708, 155710, 155712-155713.**

Batrachoseps gregarius Jockusch, Wake, and Yanev

Contrib. Sci., Nat. Hist. Mus. Los Angeles County 472:2. 1998.

HOLOTYPE: **MVZ 224581** (adult female); "Westfall Picnic Ground east of Highway 41, Sierra National Forest, Madera-Mariposa county line, California,

[USA,] 37° 26′ 43″ N, 119° 39′ 01″ W, 1400 m elevation"; E. L. Jockusch (1178) and G. Parra-Olea; April 3, 1995.

PARATOPOTYPES: **MVZ 158210-158212, 224551, 224557-224558, 224563, 224576, 224578, 224591-224594, 224596, 224606, 224611, 224614.**

Batrachoseps incognitus Jockusch, Yanev, and Wake

Herpetol. Monogr. 15:67. 2001.

HOLOTYPE: **MVZ 100059** (adult male); "near Rocky Butte, 14.7 km NE [of] Highway 1 on San Simeon Creek Road, San Luis Obispo County, CA [=California], [USA,] (ca. 35° 41′ N, 121° 03′ W)[,] [e]levation ca. 900 m"; D. B. Wake and K. Yanev; January 22, 1972.

PARATOPOTYPES: **MVZ 100047, 100049, 100051-100052, 100058, 100062, 100071, 100073, 100075-100078, 100084-100085.**

PARATYPES: **MVZ 224787-224790, 224792**; east slope of Pine Mountain, 22.9 km NE of Highway 1 on San Simeon Creek Road, San Luis Obispo County, California, USA, ca. 900 m elevation (35.683° N, 121.076° W).

Batrachoseps kawia Jockusch, Wake, and Yanev

Contrib. Sci., Nat. Hist. Mus. Los Angeles County 472:11. 1998.

HOLOTYPE: **MVZ 94134** (adult male); "west side of the South Fork, Kaweah River, Tulare County, California, [USA,] 36° 22′ 57″ N, 118° 52′ 15″ W, ca 430 m elevation"; J. L. Edwards, S. S. Sweet, D. B Wake, and R. Wassersug; December 6, 1970.

PARATOPOTYPES: **MVZ 94126-94130, 94133, 94135-94137, 94139, 94141, 94144-94145, 94148, 94152-94153.**

Batrachoseps luciae Jockusch, Yanev, and Wake

Herpetol. Monogr. 15:61. 2001.

HOLOTYPE: **MVZ 224757** (adult female); "Don Dahvee Park, Monterey, Monterey Co., CA [=California], [USA,] 36° 35′ 22″ N, 121° 53′ 38″ W, [e]levation ca. 24 m"; E. L. Jockusch (654); November 12, 1993. Field notes of E. L. Jockusch in the MVZ list S. M. Deban, M. García-París and G. Parra-Olea as the other collectors of MVZ 224757.

PARATOPOTYPES: **MVZ 104941, 104953, 104963, 104976, 104985, 224730-224732, 224735, 224737, 224758, 224760-224764, 224766, 224782-224783.**

Batrachoseps major Camp

Univ. California Publ. Zool. 12:327. 1915.

HOLOTYPE: **MVZ 611** (adult female); "Sierra Madre, 1000 feet altitude, Los Angeles County, California[, USA]"; C. L. Camp (218); March 14, 1909.

Batrachoseps minor Jockusch, Yanev, and Wake

Herpetol. Monogr. 15:65. 2001.
HOLOTYPE: **MVZ 155968** (adult male); "along Santa Rita-Old Creek Road, 8.5 km SW [of] intersection with Vineyard Road, San Luis Obispo County, California, [USA,] ca. 400 m [elevation], 35° 31′ 20″ N, 120° 46′ 40″ W"; K. Yanev (4012), S. S. Sweet, and A. Greer; March 11, 1975. Field notes of K. Yanev in the MVZ list P. Greer as one of the collectors of MVZ 155968.
PARATOPOTYPES: **MVZ 155937-155938, 155947-155948, 155951-155954, 155956, 155962-155967, 155969-155972.**

Batrachoseps regius Jockusch, Wake, and Yanev

Contrib. Sci., Nat. Hist. Mus. Los Angeles County 472:9. 1998.
HOLOTYPE: **MVZ 94029** (adult male); "south bank of the North Fork, Kings River, 1.6 km N (by road) of the Kings River, Fresno County, California, [USA,] 36° 52′ 46″ N, 119° 07′ 30″ W, ca 335 m elevation"; J. W. Crim, J. L. Edwards, S. B. Ruth, and S. S. Sweet; December 5, 1970.
PARATOPOTYPES: **MVZ 94022-94023, 94025-94026, 94030-94032, 94034, 94036, 94038-94039, 94047-94048, 94056-94057, 224799-224800, 224803.**

Batrachoseps relictus Brame and Murray

Bull. Nat. Hist. Mus. Los Angeles County 4:5. 1968.
PARATYPES: **MVZ 63837, 63839-63840, 63842-63845**; Cow Flat Creek at Highway 178, 3.4 mi SW of Democrat Hot Springs Resort turnoff, 22.3 mi (by road) NE of Bakersfield, Kern County, California, USA, 1,800 ft elevation. **MVZ 184906-184908**; 1.2 mi S of White River Lower Camp Grounds, Tulare County, California, USA.
REMARKS: MVZ 184906-184908 are cleared and stained skeletal preparations and are referred to as "3 cleared & stained specimens in D. B. Wake collection" in the original description (Brame and Murray, 1968:6).

Batrachoseps simatus Brame and Murray

Bull. Nat. Hist. Mus. Los Angeles County 4:15. 1968.
PARATYPES: **MVZ 56585-56589**; Cow Flat Creek, ca. 3-3.5 mi SW of Democrat Hot Springs Resort turnoff, south side of the Kern River Canyon above Highway 178, ca. 20 mi ENE of Bakersfield, Kern County, California, USA, 1,800 ft elevation.

Batrachoseps stebbinsi Brame and Murray

Bull. Nat. Hist. Mus. Los Angeles County 4:18. 1968.
HOLOTYPE: **MVZ 81835** (adult female); "3 mi[les] west of Paris Loraine (sometimes called Loraine), Piute Mountains, southern Sierra Nevada, Kern County, California, [USA,] at 2500 feet elevation, beneath rock tallus [sic] along west side of stream"; J. M. Brode (R. C. Stebbins 11619); April 22, 1967.
PARATOPOTYPES: **MVZ 81832-81834**.
PARATYPE: **MVZ 65773**; 6.3 mi SE of Keene Store (Post Office) on Highway 466, at the base of the northern slopes of Black Mountain, Tehachapi Mountains, Kern County, California, USA, ca. 3,000 ft elevation.
REMARKS: MVZ 81834 is a cleared and stained skeletal preparation.

Bolitoglossa celaque McCranie and Wilson

Herpetologica 49:11. 1993.
PARATYPES: **MVZ 186728-186731, 186733-186739, 186741-186742, 186745-186746**; eastern slope of Cerro Celaque, Cordillera de Celaque, Departamento Lempira, Honduras, 1,930-2,620 m elevation. **MVZ 186751**; Zacate Blanco, Departamento Intibucá, Honduras, 2,070 m elevation.

Bolitoglossa conanti McCranie and Wilson

Herpetologica 49:4. 1993.
PARATYPES: **MVZ 128271-128272, 132949-132950**; mountains W of San Pedro Sula, Departamento Cortés, Honduras, ca. 1,500-1,570 m elevation. **MVZ 186757-186758, 186760**; Cerro Cusuco, Departamento Cortés, Honduras, 1,540-1,560 m elevation. **MVZ 186762-186763, 186767, 186769, 186777**; Quebrada Grande, Departamento Copán, Honduras, 1,370-1,680 m elevation.
REMARKS: MVZ 13249-13250 (*Callisaurus draconoides*) were erroneously listed as paratypes in the original description. MVZ 132949-132950 match the data presented by McCranie and Wilson and therefore are believed to be paratypes. In addition, loan records in the MVZ show that Wilson borrowed the latter specimens, not MVZ 13249-13250.

Bolitoglossa diaphora McCranie and Wilson

J. Herpetol. 29:448. 1995.
HOLOTYPE: **MVZ 221178** (adult male); "above the visitors center of Parque Nacional El Cusuco, Cerro Cusuco (15° 31' N, 88° 12' W), 5.6 km WSW [of] Buenos Aires, 1550 m elevation, Sierra de Omoa, Departamento Cortés, Honduras"; J. R. McCranie (9688); August 21, 1992.
PARATOPOTYPES: **MVZ 186764, 221180**.
PARATYPES: **MVZ 221179**; Cerro Jilinco, Departamento Cortés, Honduras, 2,200 m elevation.

Bolitoglossa digitigrada Wake, Brame, and Thomas

Occ. Pap. Mus. Zool., Louisiana State Univ. 58:3. 1982.
PARATOPOTYPES: **MVZ 175848-175849**; "along the Río Santa Rosa[,] a few kilometers upstream from the Río Apurímac[,] between Pataccocha and San José (12° 44' S, 73° 46' W), Departamento de Ayacucho, Perú, at an elevation of 1000 m (3300 ft.)."
REMARKS: MVZ 175848-175849 were formerly LSUMZ 25515 and 25513, respectively. MVZ 175849 is a cleared and stained skeletal preparation.

Bolitoglossa gracilis Bolaños, Robinson, and Wake

Rev. Biol. Trop. 35:87. 1987.
PARATOPOTYPES: **MVZ 194888, 200853**; "Río Quirí, about 1 km NE by road from the bridge crossing the Río Grande de Orosi near Tapantí, Cartago Province, Costa Rica, at an elevation of approximately 1,280 m. The exact location is [at] 9° 47' 30" N and 83° 47' 42" W."
PARATYPE: **MVZ 200854**; vicinity of Tapantí, Provincia Cartago, Costa Rica.
REMARKS: MVZ 200853-200854 were formerly UCR 4944 and 9377, respectively.

Bolitoglossa hartwegi Wake and Brame

Contrib. Sci., Nat. Hist. Mus. Los Angeles County 175:10. 1969.
PARATYPES: **MVZ 57098-57115**; 6 mi SE of San Cristóbal de las Casas, Chiapas, México, 7,300 ft (2,226 m) elevation. **MVZ 57737-57738**; 32 mi SE of San Cristóbal de las Casas, Chiapas, México, 7,500 ft (2,287 m) elevation. **MVZ 66191**; 35 mi SE of San Cristóbal de las Casas, Chiapas, México, 7,000 ft (2,134 m) elevation.

Bolitoglossa hermosa Papenfuss, Wake, and Adler

J. Herpetol. 17:295. 1984 (1983).
HOLOTYPE: **MVZ 143804** (adult female); "4.2 km (by road) E of Rio Santiago, elevation 825 m, Guerrero, Mexico"; T. J. Papenfuss and R. L. Seib; December 23, 1976.
PARATYPES: **MVZ 143805-143808, 158489-158506, 177900**; 18.1 km NE (by Puerto del Gallo Road) of Atoyac, Guerrero, México, 775 m elevation.
REMARKS: MVZ 158491, 158493, and 177900 are cleared and stained skeletal preparations.

Bolitoglossa jacksoni Elias

Contrib. Sci., Nat. Hist. Mus. Los Angeles County 348:7. 1984.

HOLOTYPE: **MVZ 134634** (adult female); "Las Nubes sector of Finca Chiblac, approximately 12 km NNE of Santa Cruz Barillas, Depto. [=Departamento] Huehuetenango, Guatemala, at about 1,400 m elevation"; J. L. Jackson (P. Elias 371); September 1, 1975.

Bolitoglossa longissima McCranie and Cruz

Carib. J. Sci. 32:195. 1996.

HOLOTYPE: **MVZ 222552** (adult female); "along the trail to Pico La Picucha, Sierra de Agalta (14° 58′ N, 85° 55′ W), ca. 10 airline km NNW [of] Catacamas, 1900 m elevation, Departamento de Olancho, Honduras"; G. A. Cruz; August 15, 1991.

REMARKS: MVZ 222552 was formerly UNAH 2596.

Bolitoglossa meliana Wake and Lynch

Herpetologica 38:258. 1982.

HOLOTYPE: **MVZ 160736** (adult female); "vicinity of Santa Rosa Pass, 9 km NE [of] Santa Cruz del Quiché, [Departamento] El Quiché, Guatemala (elevation 2520 m)"; P. Elias, E. J. Koford, D. B. Wake, and T. A. Wake; July 16, 1978.

PARATOPOTYPES: **MVZ 160361-160399, 160737-160771.**

PARATYPES: **MVZ 108854, 113160-113161, 150519, 169069**; 4-6 km (by road) S of Purulhá, Departamento Baja Verapaz, Guatemala, 1,650-1,800 m elevation. **MVZ 150804-150807, 150813**; San Antonio, 8 km (by road) ESE of Chilascó, Departamento Baja Verapaz, Guatemala, 1,850 m elevation. **MVZ 150789, 150812**; Finca Planada, 15 km NNE of Río Hondo, Departamento Zacapa, Guatemala, 1,700 m elevation. **MVZ 150808-150811, 160772**; Sierra de las Minas, 12 km N of Santa Cruz, Departamento Zacapa, Guatemala, 2,200 m elevation. **MVZ 169038-169039**; La Bella crest, 20 km NNW of San Agustín Acasaguastlán, Departamento El Progreso, Guatemala, 2,725 m elevation.

REMARKS: MVZ 160740 is a cleared and stained skeletal preparation.

Bolitoglossa synoria McCranie and Köhler

Senckenberg. Biol. 78:226. 1999.

PARATYPES: **MVZ 39733**; "Los Esesmiles," northwest slope of Cerro El Pital, Departamento Chalatenango, El Salvador, ca. 2,438 m elevation. **MVZ 39736, 39739**; same locality as MVZ 39733, ca. 2,713 m elevation. **MVZ 39743-39744**; "Los Esesmiles," east slope of Cerro El Pital, Departamento Chalatenango, El Salvador, ca. 2,469 m elevation. **MVZ 39761, 39765, 39767, 39770, 39772**; same locality as MVZ 39743-39744, ca. 2,652 m elevation.

Bolitoglossa walkeri Brame and Wake

Contrib. Sci., Nat. Hist. Mus. Los Angeles County 219:18. 1971.
PARATYPES: **MVZ 68627-68628**; 4 km NW of San Antonio, Departamento Valle del Cauca, Colombia, 1,982 m (6,500 ft) elevation.

Bradytriton silus Wake and Elias

Contrib. Sci., Nat. Hist. Mus. Los Angeles County 345:4. 1983.
HOLOTYPE: **MVZ 131587** (adult female); "Finca Chiblac, 15 km NE [of] Barillas, Depto. [=Departamento] Huehuetenango, Guatemala, elevation 1,310 m (4,300 ft)"; P. Elias (146); September 6, 1974.
PARATOPOTYPES: **MVZ 131586, 131588-131594.**
PARATYPES: **MVZ 134635-134638**; El Rayo, 3 km S of buildings of Finca Chiblac, 10 km NE of Barillas, Departamento Huehuetenango, Guatemala, 4,500 ft (1,370 m) elevation. **MVZ 173063-173064**; Finca Chiblac, 10 km NE of Barillas, Departamento Huehuetenango, Guatemala, 4,500 ft (1,370 m) elevation.
REMARKS: MVZ 173063-173064 are cleared and stained skeletal preparations.

Chiropterotriton cuchumatanus Lynch and Wake

Contrib. Sci., Nat. Hist. Mus. Los Angeles County 265:6. 1975.
CURRENT NAME: *Dendrotriton cuchumatanus* (Lynch and Wake)
HOLOTYPE: **MVZ 113002** (adult female); "forest along [H]ighway 9N, 8.5 km (by road) SW [of] San Juan Ixcoy, [Departamento] Huehuetenango, Guatemala, at about 2860 m elev [=elevation]"; J. F. Lynch (G-7027), D. B. Wake, L. D. Houck, and A. B. Bennett; July 14, 1973.
PARATOPOTYPES: **MVZ 113003-113005, 113007-113019, 113021-113022.**
REMARKS: MVZ 113013 is a cleared and stained skeletal preparation.

Chiropterotriton rabbi Lynch and Wake

Contrib. Sci., Nat. Hist. Mus. Los Angeles County 265:2. 1975.
CURRENT NAME: *Dendrotriton rabbi* (Lynch and Wake)
HOLOTYPE: **MVZ 103839** (adult male); "9.5 km W, 8.5 km S (airline) [of] La Democracia, [Departamento] Huehuetenango, Guatemala[.] The holotype is one of a series taken at elevations between 2100 and 2500 meters"; J. F. Lynch (G-1752) and D. Bradford; December 3, 1971.
PARATOPOTYPES: **MVZ 103840-103847, 103849-103852, 103854-103871, 103873-103878, 109297-109301.**
REMARKS: MVZ 103850, 103865, 103868, 103870, 103875, and 103878 are cleared and stained skeletal preparations.

Chiropterotriton veraepacis Lynch and Wake

Contrib. Sci., Nat. Hist. Mus. Los Angeles County 294:2. 1978.
CURRENT NAME: *Cryptotriton veraepacis* (Lynch and Wake)
HOLOTYPE: **MVZ 112499** (adult male); "4.2 km (by road) S [of] Purulhá, [Departamento] Baja Verapaz, Guatemala[.] The holotype is one of a series taken at elevations between 1740 and 1780 meters"; J. F. Lynch (G-9725), D. B. Wake, and T. J. Papenfuss; November 14, 1974.
PARATOPOTYPES: **MVZ 112490-112498.**
REMARKS: MVZ 112492 is a cleared and stained skeletal preparation.

Ensatina eschscholtzii picta Wood

Univ. California Publ. Zool. 42:425. 1940.
HOLOTYPE: **MVZ 27471** (adult male); "Klamath, Del Norte County, California[, USA]"; W. F. Wood (33-1704); November 4, 1933.
PARATOPOTYPE: **MVZ 27472.**

Ensatina eschscholtzii xanthoptica Stebbins

Univ. California Publ. Zool. 48:407. 1949.
HOLOTYPE: **MVZ 41726** (adult male); "4.5 miles east [of] Schellville, Napa County, California[, USA]"; R. C. Stebbins (390); November 25, 1945.

Ensatina sierrae Storer

Univ. California Publ. Zool. 30:448. 1929.
CURRENT NAME: *Ensatina eschscholtzii platensis* (Espada)
HOLOTYPE: **MVZ 10202** (male); "Yosemite Valley, Mariposa County, California[, USA,] actually at 7300 feet [elevation], near Dewey Point, on south rim of valley"; L. K. Wilson; July 11, 1925.

Eurycea naufragia Chippindale, Price, Wiens, and Hillis

Herpetol. Monogr. 14:37. 2000.
PARATYPE: **MVZ 122775**; Montell Creek Spring, 5.2 mi (by air) NW of Montell, Uvalde County, Texas, USA, 1,530 ft elevation.

Eurycea tonkawae Chippindale, Price, Wiens, and Hillis

Herpetol. Monogr. 14:32. 2000.
PARATOPOTYPES: **MVZ 122695-122703**; "primary outflows of Stillhouse Hollow Springs, [Bull Creek watershed] Travis Co.[,] Texas, [USA, 800 feet elevation,] 30° 22' 28" N, 97° 45' 55" W." The type locality is 6.8 mi (by air) NNW of the state capital in Austin.

Hydromantes brunus Gorman

Herpetologica 10:153. 1954.
HOLOTYPE: **MVZ 59530** (female); "base of low cliffs beside State Route 140, 0.7 miles NNE [of] Briceburg (confluence of Bear Creek and Merced River), Mariposa County, California[, USA]. The elevation at the roadside is 1,285 feet"; G. Gorman (J. Gorman 434); February 24, 1952.
PARATYPE: **MVZ 55939**; hillside above the type locality, ca. 1,450 ft elevation.

Hydromantes shastae Gorman and Camp

Copeia 1953:39. 1953.
HOLOTYPE: **MVZ 52314** (adult female); "entrance to limestone caves at the edge of Flat Creek Road in the narrows of Low Pass Creek (0.7 mi. east of Squaw Creek Road, 18.4 mi. north and 15.3 mi. east of Redding), Shasta County, California[, USA]. Elevation 1500 ft."; J. Gorman (174); June 12, 1950.
PARATOPOTYPE: **MVZ 52318**; about 35 m E of type locality.

Ixalotriton niger Wake and Johnson

Contrib. Sci., Nat. Hist. Mus. Los Angeles County 411:2. 1989.
HOLOTYPE: **MVZ 158823** (adult female); "12 km (7.5 mi) NW [of] Berriozábal, Chiapas, Mexico, elevation ca. 1,068 m"; J. D. Johnson; December 26, 1980.
PARATOPOTYPES: **MVZ 143837-143838, 158822, 160952-160958, 160960-160963, 184891-184892.**
REMARKS: MVZ 184891-184892 are cleared and stained skeletal preparations.

Nototriton adelos Papenfuss and Wake

Acta Zool. Mex. (Nueva Ser.) 21:7. 1987.
CURRENT NAME: *Cryptotriton adelos* (Papenfuss and Wake)
HOLOTYPE: **MVZ 112226** (adult male); "65 km NE (by Mex. Hwy. [=México Highway] 175) [of] Guelatao, Oaxaca, Mexico"; T. J. Papenfuss (10934); August 22, 1974.
PARATOPOTYPE: **MVZ 112225**.
PARATYPES: **MVZ 183359**; 0.5 km S of Vista Hermosa (along México Highway 175), Oaxaca, México, 1,530 m elevation. **MVZ 208582**; La Esperanza, Tuxtepec, Oaxaca, México.
REMARKS: MVZ 208582 was referred to as "University of Rome (uncatalogued)" in the original description and was received as a gift from the University of Rome I "La Sapienza," Rome, Italy.

Nototriton alvarezdeltoroi Papenfuss and Wake

Acta Zool. Mex. (Nueva Ser.) 21:9. 1987.
CURRENT NAME: *Cryptotriton alvarezdeltoroi* (Papenfuss and Wake)
HOLOTYPE: **MVZ 158942** (adult male); "21.5 miles (34.6 km) N (by Mex. Hwy. [=México Highway] 195) of Jitotol, Chiapas, Mexico, ca. 1550 m (5100 feet) [elevation]"; T. J. Papenfuss (13447) and R. L. Seib; August 14, 1976.
PARATYPE: **MVZ 201396**; Puerto del Viento, 13 km NW of Pueblo Nuevo Solistahuacán, Chiapas, México, 6,000 ft (1,829 m) elevation.
REMARKS: Papenfuss and Wake (1987:9) stated that they believe MVZ 201396 was collected at the type locality, even though the elevation was recorded as 6,000 ft. If this is correct, MVZ 201396 is a paratopotype.

Nototriton gamezi García-París and Wake

Copeia 2000:55. 2000.
HOLOTYPE: **MVZ 207122** (adult female); "Carril Bosque Eterno at junction with Pantanosa Trail, Monteverde Cloud Forest Reserve, Prov. [=Provincia] Alajuela, Costa Rica, elev. [=elevation] 1600 m, approximately 10° 19' N, 84° 47.5' W"; D. C. Cannatella, D. A. Good (3207), W. Guindon, and D. B. Wake; August 14, 1987.
PARATYPES: **MVZ 207120-207121, 207123**; Peñas Blancas trail below (E) Continental Divide, Monteverde Cloud Forest Reserve, Provincia Alajuela, Costa Rica, 1,530-1,540 m elevation. **MVZ 207124**; Pantanosa Trail, Monteverde Cloud Forest Reserve, Provincia Alajuela, Costa Rica, 1,590 m elevation.

Nototriton guanacaste Good and Wake

Herpetol. Monogr. 7:138. 1993.

HOLOTYPE: **MVZ 207111** (adult female); "upper slopes of Volcán Cacao (elevation 1580 m), Guanacaste National Park, Prov. [=Provincia] de Guanacaste, Costa Rica"; D. C. Cannatella and D. A. Good (3292); August 24, 1987.

PARATOPOTYPES: **MVZ 207106-207110, 207112-207114.**

REMARKS: MVZ 207106 is a cleared and stained skeletal preparation.

Nototriton limnospectator McCranie, Wilson, and Polisar

Herpetologica 54:455. 1998.

PARATYPE: **MVZ 225866**; El Ocotillo, northeastern slope of Montaña de Santa Bárbara, Departamento Santa Bárbara, Honduras, 1,800 m elevation (14° 56' N, 88° 06' W).

Nototriton tapanti Good and Wake

Herpetol. Monogr. 7:134. 1993.

HOLOTYPE: **MVZ 203746** (adult female); "Rio Quirí (elevation ca. 1300 m), 0.3 km northeast of the junction of the Tapantí Road and the Tausito Road, near Tapantí, Prov. [=Provincia] Cartago, Costa Rica"; A. Collazo and D. A. Good (T-192); May 25, 1986.

Nyctanolis pernix Elias and Wake

Advances in Herpetology and Evolutionary Biology: Essays in Honor of Ernest E. Williams (Rhodin and Miyata, eds.). Mus. Comp. Zool., Cambridge, Massachusetts, p. 2. 1983.

HOLOTYPE: **MVZ 134641** (adult female); "Finca Chiblac, 10 km ([by] air) NE [of] Barillas, [Departamento] Huehuetenango, Guatemala, (91° 16' W, 15° 53' N), 1,370 m (4,500 ft) elevation"; P. Elias (311) and J. Jackson; August 29, 1975.

PARATOPOTYPES: **MVZ 131583-131585, 134639-134640, 134642-134644, 149370, 149372-149373, 173062.**

REMARKS: MVZ 173062 is a cleared and stained skeletal preparation.

Oedipina maritima García-París and Wake

Copeia 2000:50. 2000.
PARATOPOTYPE: **MVZ 219997**; "Escudo Camp, West Point [=west end of] Isla Escudo de Veraguas, Prov. [=Provincia] Bocas del Toro, Panamá, approximate 9° 6.1′ N, 81° 4.5′ W."
REMARKS: MVZ 219997 was formerly USNM Field Herp 195520.

Oedipina savagei García-París and Wake

Copeia 2000:52. 2000.
PARATYPE: **MVZ 229360**; Paraguas Ridge (slopes of Cerro Zapote), ca. 4 km W of Agua Buena, Provincia Puntarenas, Costa Rica, ca. 1,400 m elevation (ca. 08° 45′ N, 82° 59′ W).

Plethodon longicrus Adler and Dennis

Ohio Herpetol. Soc. Spec. Publ. 4:1. 1962.
CURRENT NAME: *Plethodon yonahlossee* Dunn
PARATOPOTYPES: **MVZ 72661-72662**; "northeast slope of Bluerock Mountain, below the Bat Caves, Rutherford County, North Carolina[, USA], approximately 0.8 mile ESE of Bat Cave ([C]ity), at an elevation of about 1645 feet about sea level (data according to USGS map, Bat Cave quadrangle, 1946 edition)."

Plethodon neomexicanus Stebbins and Riemer

Copeia 1950:73. 1950.
HOLOTYPE: **MVZ 49033** (adult male); "12 miles west and 4 miles south of Los Alamos, 8750 ± feet [elevation], Sandoval County, New Mexico[, USA]"; R. C. Stebbins (3447); August 14, 1949.
PARATOPOTYPES: **MVZ 49018-49032, 49034-49035**.
REMARKS: MVZ 49027 was sent on permanent loan to UMMZ and now is UMMZ 109902. MVZ 49035 was sent on permanent loan to USNM and now is USNM 129378. MVZ 49024 and 49032 are cleared and stained skeletal preparations but were erroneously listed as skeletons in the original description. MVZ 49030 and 49034 are missing.

Plethodon richmondi popei Highton and Grobman

Herpetologica 12:187. 1956.
PARATOPOTYPES: **MVZ 63723-63724**; "Comers Rock, Grayson-Wythe County line, Virginia[, USA]."

Plethodon richmondi shenandoah Highton and Worthington

Copeia 1967:619. 1967.
PARATYPES: **MVZ 79379-79380**; Appalachian Trail, 0.57 mi (by trail) W of junction with Timber Hollow Fire Trail, 0.1 mi (by air) N of top of Hawksbill Mountain, Shenandoah National Park, Page County, Virginia, USA.

Plethodon stormi Highton and Brame

Pilot Register of Zoology, Card No. 20. 1965.
PARATYPE: **MVZ 78571**; 0.5 mi N of the California state border, on Applegate River Road, Jackson County, Oregon, USA.

Pseudoeurycea aquatica Wake and Campbell

Herpetologica 57:509. 2001.
PARATOPOTYPE: **MVZ 230728**; "5.8 km (by road) W [of] Totontepec, Oaxaca, Mexico, 2103 m [elevation] (17° 14′ 24″ N, 96° 03′ 36″ W)."
REMARKS: MVZ 230728 was formerly UTA A-5834.

Pseudoeurycea juarezi Regal

Amer. Mus. Novitates 2266:1. 1966.
PARATYPES: **MVZ 194022-194024**; cloud forest on NE slope of Sierra Juárez, ca. 58 km S of Valle Nacional on road between Ixtlán de Juárez and Tuxtepec, Oaxaca, México (17° 48′ N, 96° 20′ W).
REMARKS: MVZ 194022-194024 are referred to as P. J. Regal 1095, 1096, and 1100, respectively, in the original description. MVZ 194022 and 194024 are cleared and stained skeletal preparations.

Pseudoeurycea longicauda Lynch, Wake, and Yang

Copeia 1983:887. 1983.
HOLOTYPE: **MVZ 137880** (adult female); "slope just South [of] Mex. hwy [=México Highway] 15, 23.1 km (by rd [=road]) W. [of] Villa Victoria, State of Mexico, Mexico (elevation 2,850-2,970 m)"; M. Feder, J. F. Lynch (M-4107), and D. B. Wake; July 7, 1976.
PARATOPOTYPES: **MVZ 137879, 137881-137921, 138418-138436, 138518-138519.**
REMARKS: MVZ 138518 is missing.

Pseudoeurycea lynchi Parra-Olea, Papenfuss, and Wake

Sci. Pap. Nat. Hist. Mus., Univ. Kansas 20:3. 2001.
HOLOTYPE: **MVZ 230994** (adult female); "top of Cerro San Pedro at Loma Alta Microwave Station, 6 km ENE [of] Chiconquiaco, Veracruz, Mexico (19° 45.3′ N, 96° 46.9′ W, 2160 m elevation)"; G. Parra-Olea (153), M. García-París, and D. B. Wake; October 15, 1997. The specimen tag lists part of the coordinates as "96° 00″ W," but this is in error.
PARATOPOTYPES: **MVZ 230995-231000.**
PARATYPES: **MVZ 158821, 158824-158825, 178841-178843, 203669-203670, 203672-203675**; forest west of La Joya, Veracruz, México, 2,120 m elevation (19° 36.9′ N, 97° 01.9′ W).

Pseudoeurycea mixcoatl Adler

Occ. Pap. Nat. Hist. Mus., Univ. Kansas 177:16. 1996.
PARATYPES: **MVZ 110899-110901**; 0.6 km (by road) SW of Carrizal de Bravo [=Corral de Bravo], Sierra Madre del Sur, Guerrero, México, 2,200 m elevation (17° 37′ N, 99° 50′ W). **MVZ 132881-132893, 162162-162172**; 4 km (by road) SW of Carrizal de Bravo [=Corral de Bravo], Sierra Madre del Sur, Guerrero, México, 8,460 ft (2,580 m) elevation.

Pseudoeurycea naucampatepetl Parra-Olea, Papenfuss, and Wake

Sci. Pap. Nat. Hist. Mus., Univ. Kansas 20:4. 2001.
HOLOTYPE: **MVZ 158941** (adult male); "along the road to Las Lajas Microwave Station, 15 km (by rd [=road]) S [of] Highway 140 from Las Vigas, Veracruz, Mexico (19° 35.5′ N, 97° 05.7′ W, 3000 m elevation)"; T. J. Papenfuss; October 24, 1981. The specimen tag and field notes of D. B. Wake in the MVZ indicate that the field number of MVZ 158941 is DBW 1249.
PARATOPOTYPES: **MVZ 171521, 172131, 173436.**

Pseudoeurycea parva Lynch and Wake

Contrib. Sci., Nat. Hist. Mus. Los Angeles County 411:15. 1989.
CURRENT NAME: *Ixalotriton parvus* (Lynch and Wake)
HOLOTYPE: **MVZ 196101** (adult male); "ridge SE [of] Cerro Baul, 21 km W [of] Rizo de Oro, [in] Chiapas, [Oaxaca,] Mexico (ca. 1,600 m [elevation]). The type-locality is just within the eastern border of the state of Oaxaca"; K. Lucas (M-1441); September 8, 1972.
PARATOPOTYPES: **MVZ 196102, 196444-196445, 202294.**
PARATYPES: **MVZ 163823-163825, 177823-177824**; 16.8 km (by road) NW of Rizo de Oro (in Chiapas), Oaxaca, México, 1,650-1,860 m elevation. **MVZ 194330-194331**; Cerro Baúl, 37 km (by road) W of Rizo de Oro (in Chiapas), Oaxaca, México.

REMARKS: MVZ 202294 is a cleared and stained skeletal preparation.

Pseudoeurycea saltator Lynch and Wake

Contrib. Sci., Nat. Hist. Mus. Los Angeles County 411:11. 1989.
HOLOTYPE: **MVZ 131102** (adult male); "cloud forest just west of [México] [H]ighway 175, 16 km (by road) S [of] Vista Hermosa, Oaxaca, Mexico (1,970 m [elevation])"; D. B. Wake (M-2518), J. F. Lynch, and T. J. Papenfuss; November 21, 1974.
PARATYPES: **MVZ 112227-112237, 114398, 132876-132880, 147259-147263, 162283**; within 1 km of type locality on west side of México Highway 175, Oaxaca, México, 1,950-2,050 m elevation.
REMARKS: MVZ 147260 is a cleared and stained skeletal preparation.

Pseudoeurycea tlahcuiloh Adler

Occ. Pap. Nat. Hist. Mus., Univ. Kansas 177:10. 1996.
PARATYPES: **MVZ 222469**; 29.5 km (by road) W of Cruz Ocote, along the Milpillas-Atoyac Road, on the eastern approaches to Cerro Teotepec, Sierra Madre del Sur, Guerrero, México, 8,740 ft (2,666 m) elevation. **MVZ 222470-222471**; 36.8 km (by road) W of Cruz Ocote, along the Milpillas-Atoyac Road, on the eastern approaches to Cerro Teotepec, Sierra Madre del Sur, Guerrero, México, 8,775-8,900 ft (2,676-2,715 m) elevation.

Spelerpes dofleini Werner

[McCranie, Wake, and Wilson.] Carib. J. Sci. 32:397. 1996.
CURRENT NAME: *Bolitoglossa dofleini* (Werner)
NEOTYPE: **MVZ 161627** (adult female); "Finca [El] Volcán, 25 km (by rd. [=road]) NW [of] Senahú, Depto. [=Departamento] Alta Verapaz, Guatemala, elevation 875 m"; T. J. Papenfuss (16466) and R. L. Seib; between July 16 and July 20, 1978.

Spelerpes platycephalus Camp

Univ. California Publ. Zool. 17:11. 1916[a].
CURRENT NAME: *Hydromantes platycephalus* (Camp)
HOLOTYPE: **MVZ 5693** (adult female); "head of Lyell Cañon, 10,800 feet altitude, Yosemite National Park, [Tuolumne County,] California[, USA]"; C. L. Camp (2215); July 18, 1915.
PARATOPOTYPE: **MVZ 5694**.
REMARKS: *Spelerpes platycephalus* was designated as the type species of the genus *Hydromantoides* Lanza and Vanni (Lanza and Vanni, 1981), which was established for the three North American species of *Hydromantes*.

The biological basis for generic separation of North American and European species was questioned by Jackman et al. (1997). By action of the International Commission on Zoological Nomenclature (Anonymous, 1997), *Hydromantoides* was placed on the Official Index of Rejected and Invalid Generic Names in Zoology, and *Spelerpes platycephalus* was designated the type species of *Hydromantes*.

Thorius arboreus Hanken and Wake

Copeia 1994:578. 1994.
HOLOTYPE: **MVZ 196078** (adult female); "10.9 km (rd.) [=by road] W [of] La Esperanza, along [México] Hwy. [=Highway] 175, Oaxaca, México, elev. [=elevation] 2060 m"; D. M. Darda (412) and P. A. Garvey; November 20, 1983.
PARATYPES: **MVZ 112210**; 65 km NE of Guelatao on México Highway 175, Oaxaca, México. **MVZ 131440**; 16 km (by road) S of Vista Hermosa on México Highway 175, Oaxaca, México, 2,050 m elevation. **MVZ 158915**; 65 km NE of Guelatao on México Highway 175, Oaxaca, México, 6,480 ft elevation. **MVZ 178844**; 66 km N of Guelatao on México Highway 175, Oaxaca, México, 2,100 m elevation. **MVZ 183349, 187010-187011**; México Highway 175, 33.6 mi N of Guelatao, Oaxaca, México, 2,380 m elevation. **MVZ 183350, 187012-187013**; México Highway 175, 34.7 mi (by road) N of Guelatao, Oaxaca, México, 7,660 ft elevation. **MVZ 183353**; 40.5 mi N of Guelatao on México Highway 175, Oaxaca, México.
REMARKS: MVZ 187010-187013 are cleared and stained skeletal preparations.

Thorius aureus Hanken and Wake

Copeia 1994:574. 1994.
HOLOTYPE: **MVZ 85966** (adult female); "0.7 mi (rd.) [=by road] E [of] Cerro Pelón from point where road crosses top, Oaxaca, México"; R. W. McDiarmid (3073) and R. D. Worthington; January 20, 1969.
PARATOPOTYPES: **MVZ 85956-85958, 85960-85961, 85968-85970, 85972-85973, 85975, 85978-85979**.
PARATYPES: **MVZ 112175, 112179, 112181-112182, 112184**; 52 km (by México Highway 175) NE of Guelatao, Oaxaca, México. **MVZ 114690**; Cerro Pelón, near México Highway 175, 108.9 km (by road) N of junction with Pan American Highway at Oaxaca City, Oaxaca, México. **MVZ 183332, 187179-187184, 187189-187191**; 32.4 mi N of Guelatao on México Highway 175, Oaxaca, México, 2,515 m elevation. **MVZ 187003-187009**; México Highway 175, 33.5 mi N of Guelatao, Oaxaca, México, 8,120 ft elevation. **MVZ 187185-187187**; México Highway 175, 31.8 mi N of Guelatao, 1 km N of Cerro Pelón from where road crosses top, Oaxaca, México.
REMARKS: MVZ 187003-187009, 187179-187187, and 187189-187191 are cleared and stained skeletal preparations.

Thorius boreas Hanken and Wake

Copeia 1994:581. 1994.

HOLOTYPE: **MVZ 162202** (adult male); "Llano de las Flores, along [México] Hwy. [=Highway] 175, 25-26 km NE [of] Guelatao, Oaxaca, México"; T. J. Papenfuss (11959); August 12, 1975.

PARATOPOTYPES: **MVZ 162188-162198, 162203-162204, 162244, 162250.**

PARATYPES: **MVZ 85985, 85995**; 11.7 mi SW of Cerro Pelón on Tuxtepec-Oaxaca road, 17 mi NE of Ixtlán de Juárez, Oaxaca, México. **MVZ 112169**; 22 km N of Guelatao along México Highway 175, Oaxaca, México. **MVZ 131363**; 33 km (by road) N of Guelatao along México Highway 175, Oaxaca, México, 2,940 m elevation. **MVZ 183327, 187162-187164**; type locality, 2,830 m elevation. **MVZ 183528, 187165-187168**; 18.9 mi (by road) N of Guelatao along México Highway 175, Oaxaca, México, 9,470 ft elevation. **MVZ 186999-187000**; 29.1 km N of Guelatao along México Highway 175, Oaxaca, México, 2,890 m elevation. **MVZ 187161**; type locality, 2,950 m elevation. **MVZ 187169-187173**; type locality, 9,340 ft elevation. **MVZ 187174-187178**; Cerro Pelón, 31.8 mi N of Guelatao, 1.0 km E from where México Highway 175 crosses top, Oaxaca, México.

REMARKS: MVZ 186999-187000 and 187161-187178 are cleared and stained skeletal preparations.

Thorius grandis Hanken, Wake, and Freeman

Copeia 1999:923. 1999.

HOLOTYPE: **MVZ 183384** (adult female); "Puerto del Gallo, Guerrero, México, elevation 2500 m"; T. J. Papenfuss (14555); December 22, 1976.

PARATOPOTYPES: **MVZ 183366-183368, 183381-183383, 183386-183387, 183389-183390, 183392-183393, 183401, 183404-183405, 183408-183409, 187032-187051.**

PARATYPES: **MVZ 183417-183418**; Cerro Teotepec, 9 km (by road) NE of Puerto del Gallo, Guerrero, México, 3,109 m elevation.

REMARKS: MVZ 187032-187051 are cleared and stained skeletal preparations.

Thorius infernalis Hanken, Wake, and Freeman

Copeia 1999:926. 1999.

HOLOTYPE: **MVZ 183426**; (adult male), "13.7 km northeast (road from Atoyac to Puerto del Gallo) of El Paraiso, Guerrero, México, elevation 1140 m"; J. E. Cadle (1746); November 18, 1977.

PARATOPOTYPE: **MVZ 183425.**

Thorius lunaris Hanken and Wake

Copeia 1998:318. 1998.

HOLOTYPE: **MVZ 183292** (adult female); "1.5 km by road west of Texmola, on southwestern slopes of Volcán Orizaba, Veracruz, México, elevation 2640 m"; J. Hanken (M-3014), H. B. Shaffer, and C. Cerón; February 2, 1976.

PARATOPOTYPES: **MVZ 183282, 183291, 183294-183296, 183298, 183300-183301, 183307-183309, 186963-186970, 186972-186977.**

PARATYPES: **MVZ 183218-183220, 183222-183224, 183228, 187081**; canyon ca. 1 km SE of El Berro, ca. 15 km (by air) SW of Pico de Orizaba, Veracruz, México, 2,500 m elevation. **MVZ 183313-183315, 183318-183320, 187098-187101, 187188**; forest above Xometla, Veracruz, México, 8,600-8,640 ft elevation.

REMARKS: MVZ 186963-186970, 186972-186977, 187081, 187098-187101, and 187188 are cleared and stained skeletal preparations.

Thorius magnipes Hanken and Wake

Copeia 1998:326. 1998.

HOLOTYPE: **MVZ 114514** (adult female); "4 km south of Puerto del Aire, Veracruz, México, elevation 2475 m"; J. F. Lynch (M-1633), C. Kimmel, and J. Kezer; January 21, 1974.

PARATOPOTYPES: **MVZ 114515-114518.**

PARATYPES: **MVZ 85948-85949, 129657-129658, 205071**; 2 mi (by road) S of Puerto del Aire, Veracruz, México. **MVZ 150541-150542, 150554-150555, 150559-150560, 150563, 150572, 150575**; 10.9 mi (by México Route 150) W of Acultzingo, then 14.8 mi east by dirt road, Veracruz, México, 2,640 m elevation. **MVZ 185392-185396, 186960-186962**; 3.2 km S of Puerto del Aire, Veracruz, México.

REMARKS: MVZ 186960-186962 and 205071 are cleared and stained skeletal preparations.

Thorius minydemus Hanken and Wake

Copeia 1998:325. 1998.

HOLOTYPE: **MVZ 131444** (adult male); "vicinity of La Joya (19° 38' 08" N, 97° 09' 13" W), Mexican Hwy. [=Highway] 140, Veracruz, México, elevation 2230 m"; J. F. Lynch (M-2613), D. B. Wake, and T. J. Papenfuss; November 24, 1974.

Thorius munificus Hanken and Wake

Copeia 1998:321. 1998.

HOLOTYPE: **MVZ 183274** (adult female); "Mexican Hwy. [=Highway] 140, 4.5 km by road west of Las Vigas, Veracruz, México, elevation 2420 m"; J. F. Lynch (M-5381), T. J. Papenfuss, and D. B. Wake; July 25, 1976.

PARATOPOTYPES: **MVZ 183255-183273, 183277, 186998.**

PARATYPES: **MVZ 183241, 183244, 183247-183248, 186980-186997;** México Highway 140 at Las Vigas, Veracruz, México, 2,525 m elevation.

REMARKS: MVZ 186980-186998 are cleared and stained skeletal preparations.

Thorius omiltemi Hanken, Wake, and Freeman

Copeia 1999:918. 1999.

HOLOTYPE: **MVZ 110916** (adult female); "3.2 km SW of Carrizal de Bravos [=Corral de Bravo], Guerrero, México, elevation 2400 m"; T. J. Papenfuss (8654); August 31, 1973.

PARATOPOTYPES: **MVZ 110913-110915, 110919, 110921-110922.**

PARATYPES: **MVZ 57116-57117, 57119;** 4.8 km W of Omiltemi, Guerrero, México, 2,500 m elevation. **MVZ 110624, 110656, 110664, 110673, 110694, 110697-110698, 110700, 110703-110704, 110706;** type locality, 2,520 m elevation. **MVZ 187014-187031, 187070;** type locality, unknown elevation. **MVZ 110902, 110906, 110908;** 0.7 km SW of Carrizal de Bravos (=Corral de Bravo), Guerrero, México, 2,200 m elevation.

REMARKS: MVZ 57117, 110624, 187014-187031, and 187070 are cleared and stained skeletal preparations.

Thorius papaloae Hanken and Wake

Herpetologica 57:516. 2001.

HOLOTYPE: **MVZ 183468** (adult female); "8 km ([by] road) NE of Concepción Pápalo, Oaxaca, México, elevation 2670 m"; J. Hanken (M-3823) and H. B. Shaffer; February 15, 1976. The specimen tag lists only J. Hanken as the collector.

PARATOPOTYPES: **MVZ 183470, 183473, 183475, 187052-187069.**

PARATYPES: **MVZ 183479, 183483, 183485-183486, 183493-183494, 183496, 183504-183505, 183508, 183510;** 11 km (by road) NE of Concepción Pápalo, Oaxaca, México, 2,820 m elevation. **MVZ 183518;** 15 km (by road) NE of Concepción Pápalo, Oaxaca, México, 2,850 m elevation.

REMARKS: As noted by Hanken and Wake (2001:520), the specimen tags of the type series misidentify the village of Concepción Pápalo as "Santos Reyes Pápalo," a different village that lies a few kilometers away. MVZ 187052-187069 are cleared and stained skeletal preparations.

Thorius smithi Hanken and Wake

Copeia 1994:583. 1994.
HOLOTYPE: **MVZ 150590** (female); "0.5 mi (rd.) [=by road] SW [of] Vista Hermosa along [México] Hwy. [=Highway] 175, Oaxaca, México"; J. E. Cadle (1136); July 14, 1977.

Thorius spilogaster Hanken and Wake

Copeia 1998:314. 1998.
HOLOTYPE: **MVZ 183109** (adult female); "forest above Xometla, Veracruz, México, elevation 8600-8640 ft"; J. F. Lynch (M-5203); July 23, 1976. The original description lists J. Hanken, D. Eakins, and C. Cerón as the collectors.
PARATOPOTYPES: **MVZ 183097, 183100, 183104-183107, 183114, 183116-183117, 187091-187097, 187102.**
PARATYPES: **MVZ 114606, 114609, 114612, 114614-114615, 114618, 114624, 114631;** type locality, 2,710-2,725 m elevation. **MVZ 183085-183087, 183089-183090, 183093, 187071-187080, 187082-187090;** canyon ca. 1 km SE of El Berro, ca. 15 km (by air) SW of Pico de Orizaba, Veracruz, México, 2,500 m elevation.
REMARKS: MVZ 187071-187080, 187082-187097, and 187102 are cleared and stained skeletal preparations.

Rhyacotritonidae

Rhyacotriton cascadae Good and Wake

Univ. California Publ. Zool. 126:15. 1992.
HOLOTYPE: **MVZ 90795** (adult female); "base of Wahkeena Falls, Multnomah County, Oregon[, USA]"; D. B. Wake (A-446), T. H. Wake, and J. Everett; December 22, 1965.
PARATOPOTYPES: **MVZ 90793-90794, 90796, 90798-90799, 90808, 90812-90813, 90815, 90819, 90824-90825.**
REMARKS: MVZ 90808 and 90812 are larvae.

Rhyacotriton kezeri Good and Wake

Univ. California Publ. Zool. 126:16. 1992.
HOLOTYPE: **MVZ 197300** (adult female); "junction of Highway 26 and Luukinen Road (at the Nehalim [=Nehalem] River Bridge), Clatsop County, Oregon[, USA]"; D. A. Good (1498); May 3, 1985.
PARATOPOTYPES: **MVZ 197301-197310.**
REMARKS: MVZ 197306 and 197308-197309 are larvae.

Rhyacotriton olympicus variegatus Stebbins and Lowe

Univ. California Publ. Zool. 50:471. 1951.
CURRENT NAME: *Rhyacotriton variegatus* Stebbins and Lowe
HOLOTYPE: **MVZ 45868** (adult female); "1.3 miles west of Burnt Ranch
Post Office, Trinity County, California[, USA]"; R. C. Stebbins (1838);
November 16, 1947.

Salamandridae

Cynops orphicus Risch

Alytes 2:46. 1983.
HOLOTYPE: **MVZ 22474** (adult male); "Dayang (=Tai-Yong[, Jiexi
County]), Shantou Region, Guangdong Province, China, 23° 35′ N, 115° 51′ E,
altitude 640 m"; J. L. Gressitt (1363); August 4, 1936.
PARATOPOTYPES: **MVZ 22416-22473, 22475-22506, 24134-24136.**
REMARKS: MVZ 22425 was sent on permanent loan to CIB, but our
efforts to locate this specimen in that collection were unsuccessful.
MVZ 22465 is a cleared and stained skeletal preparation.   MVZ 22438
is missing.

Rhithrotriton derjugini microspilotus Nesterov

Ann. Mus. Zool. Acad. Sci. Petrograd 21:2. 1916.
CURRENT NAME: *Neurergus microspilotus* (Nesterov)
SYNTYPES: **MVZ 99385-99386, 198743**; "montibus Auromana"
(=Avroman Mountains, Kurdistan Province, Iran); May 1914.   The type
locality was given as "Avroman-Dagh [Balch, Iraq and Tawale, Kordestan (*sic*)
Prov., Iran]" by Leviton et al. (1992:139), following references to more specific
localities given by Nesterov later in the article (page 11).
REMARKS: According to Leviton et al. (1992:139), the original name of
this taxon was "*Rhithrotriton derjugini* var. *microspilotus*," but Nesterov
designated the name as written by us.   Leviton et al.   (1992:183) cited the
journal as "Ann. Mus. Zool. Acad. Imp. Sci. St. Petersbourg" because they
chose "to use the full, formal journal name for the series," as the name often
changed from volume to volume or even issue to issue (K. Adler, pers. com.).
Leviton et al. (1992:139) also stated, "Nesterov is variously cited as 1916 and
1917."   Adler checked an original in the library at Cornell University [Ithaca,
New York, USA] and reported that "Nesterov's article was published in Sept.
1916 according to the original wrappers on Cornell's copy" (Adler to Leviton,
3 July 1991).   MVZ 99385-99386 and 198743 were received as part of an
exchange of material with the Zoological Institute, USSR Academy of
Sciences, Leningrad (currently the Zoological Institute, Russian Academy of
Sciences, St. Petersburg).

Triturus granulosus mazamae Myers

[G. Myers.] Copeia 1942:80. 1942[a].
CURRENT NAME: *Taricha granulosa mazamae* (Myers)
PARATYPES: **MVZ 37364-37367**; Eagle Cove, Crater Lake, Crater Lake National Park, Klamath County, Oregon, USA, 6,166 ft elevation.
REMARKS: Myers designated as paratypes of *Triturus granulosus mazamae* five then uncataloged specimens in the MVZ, but we found only four (MVZ 37364-37367) specimens in the collection that match the data presented by Myers. These four specimens are believed to be paratypes.

Triturus rivularis Twitty

Copeia 1935:73. 1935.
CURRENT NAME: *Taricha rivularis* (Twitty)
HOLOTYPE: **MVZ 18131** (male); "Gibson Creek, about one mile west of Ukiah, [Mendocino County,] California, [USA]"; V. C. Twitty; February 22, 1935.
PARATOPOTYPES: **MVZ 18132-18148**.
REMARKS: MVZ 18138 is a cleared and stained skeletal preparation.

Triturus similans Twitty

Copeia 1935:76. 1935.
CURRENT NAME: *Taricha granulosa granulosa* (Skilton)
HOLOTYPE: **MVZ 18149** (male); "Robinson Creek, Ukiah, [Mendocino County,] California, [USA]"; V. C. Twitty; April 20, 1935.
PARATOPOTYPES: **MVZ 18150-18177**.
REMARKS: MVZ 18172 is a cleared and stained skeletal preparation.

GYMNOPHIONA

Scolecomorphidae

Scolecomorphus uluguruensis Barbour and Loveridge

Mem. Mus. Comp. Zool., Harvard College 50:180. 1928.
PARATOPOTYPE: **MVZ 32982**; "Nyingwa, Uluguru Mountains, [Morogoro Region,] Tanganyika Territory [=Tanzania]."
REMARKS: MVZ 32982 was formerly MCZ 12279.

REPTILIA

SQUAMATA

Agamidae

Japalura brevipes Gressitt

Proc. Biol. Soc. Washington 49:117. 1936.
PARATOPOTYPE: **MVZ 23357**; "Bukai [=Wuchie], near Horisha [=Puli], central Formosa [=Taiwan, "Republic of China"], alt. [=altitude] 1,200 meters."

Amphisbaenidae

Blanus tingitanus Busack

Copeia 1988:106. 1988.
HOLOTYPE: **MVZ 178095** (adult male); "vicinity of the old fortress at Mehdiya-Plage, Kenitra Prefecture, Morocco"; S. D. Busack (1147) and J. A. Visnaw; April 19, 1982.
PARATOPOTYPES: **MVZ 178096-178100, 178102, 186166-186170**.

Anguidae

Abronia gaiophantasma Campbell and Frost

Bull. Amer. Mus. Nat. Hist. 216:19. 1993.
PARATYPES: **MVZ 143461**; near Chilascó, Departamento Baja Verapaz, Guatemala.   **MVZ 144537**; 4 km ENE of Chilascó, Departamento Baja Verapaz, Guatemala, 1,829 m elevation. **MVZ 160608**; Finca San Jorge, 5 km ENE of Chilascó, Departamento Baja Verapaz, Guatemala, 1,829 m elevation. **MVZ 160609**; Finca Miranda, 8 km ESE of Chilascó, Departamento Baja Verapaz, Guatemala, 1,829-1,981 m elevation.

Abronia kalaina Good and Schwenk

Copeia 1985:135. 1985.
CURRENT NAME: *Abronia fuscolabialis* (Tihen)
HOLOTYPE: **MVZ 177806** (adult male); "16.6 km (by road) north of the summit on [México] Hwy. [=Highway] 175, Cerro Pelón, Sierra Juárez, Oaxaca, México, ca 2,300 m elevation"; K. Schwenk (82) and G. Roth; March 30, 1980.

Anniella nigra argentea Hunt

Copeia 1983:86. 1983.
CURRENT NAME: *Anniella pulchra* Gray
HOLOTYPE: **MVZ 64656**; (adult male); "0.8 km S east [of] entrance, ([by] Cal. Hwy. [=California Highway] 146), to Pinnacles National Monument, San Benito County, California[, USA]"; R. C. Stebbins (7560); March 17, 1956.
PARATYPE: **MVZ 129844**; Pinnacles National Monument, San Benito County, California, USA.
REMARKS: MVZ 64656 was designated the neotype of *Anniella pulchra* Gray by Murphy and Smith (1991), a designation validated by the International Commission on Zoological Nomenclature (Anonymous, 1993).

Anniella pulchra Gray

[Murphy and Smith.] Bull. Zool. Nomencl. 48:317. 1991.
NEOTYPE: **MVZ 64656** (adult male); "0.8 km S east [of] entrance, ([by] Cal. Hwy. [=California Highway] 146), to Pinnacles National Monument, San Benito County, California[, USA]"; R. C. Stebbins (7560); March 17, 1956.
REMARKS: The petition to designate MVZ 64656 the neotype of *Anniella pulchra* (Murphy and Smith, 1991) was validated by the International Commission on Zoological Nomenclature (Anonymous, 1993).

Barisia juarezi Karges and Wright

Contrib. Sci., Nat. Hist. Mus. Los Angeles County 381:1. 1987.
CURRENT NAME: *Mesaspis juarezi* (Karges and Wright)
PARATYPES: **MVZ 112393**; 61 km NE (by México Highway 175) of Guelatao, Oaxaca, México. **MVZ 112394**; 52 km NE (by México Highway 175) of Guelatao, Oaxaca, México.

Gerrhonotus coeruleus shastensis Fitch

Copeia 1934:6. 1934[a].
CURRENT NAME: *Elgaria coerulea shastensis* (Fitch)
HOLOTYPE: **MVZ 15047** (adult male); "south side of Burney Creek, 3000 feet [elevation], two miles southwest of Burney, Shasta County, California[, USA]"; H. S. Fitch (461); June 27, 1932.

Gerrhonotus panamintinus Stebbins

Amer. Mus. Novitates 1883:2. 1958[a].
CURRENT NAME: *Elgaria panamintina* (Stebbins)
HOLOTYPE: **MVZ 65410** (adult female); "Surprise Canyon, at an elevation of 4500 feet, on the west side of the Panamint Mountains, Inyo County, California[, USA]"; J. McDonald Jr. (R. C. Stebbins 7453a); October 23, 1954.
PARATOPOTYPE: **MVZ 65407**.
PARATYPES: **MVZ 65403-65404**; Surprise Canyon, Inyo County, California, USA, ca. 4,000 ft elevation. **MVZ 65405-65406**; Brewery Spring, Panamint Mountains, Inyo County, California, USA, 4,800 ft elevation. **MVZ 65408-65409**; Limekiln Spring, Panamint Mountains, Inyo County, California, USA, 4,000 ft elevation.
REMARKS: MVZ 65404, 65406, and 65408 are shed skins.

Gerrhonotus paucicarinatus Fitch

Copeia 1934:173. 1934[b].
CURRENT NAME: *Elgaria paucicarinata* (Fitch)
HOLOTYPE: **MVZ 11768** (adult male); "Todos Santos [Island], Lower [=Baja] California, Mexico"; C. C. Lamb (9684); October 29, 1928.

Gerrhonotus scincicauda nanus Fitch

Copeia 1934:7. 1934[a].
CURRENT NAME: *Elgaria multicarinata nana* (Fitch)
HOLOTYPE: **MVZ 5402** (adult male); "South Island, Los Coronados Islands [off the coast of Tijuana], Lower [=Baja] California, Mexico"; A. B. Howell; July 1, 1913.

Eublepharidae

Coleonyx variegatus sonoriensis Klauber

Trans. San Diego Soc. Nat. Hist. 10:162. 1945.
PARATYPE: **MVZ 20839**; Ensenada del Perro, south end of Isla Tiburón, Sonora, México.

Gekkonidae

Hemidactylus stejnegeri Ota and Hikida

J. Herpetol. 23:51. 1989.
PARATOPOTYPE: **MVZ 202296**; "Hongye, Hualien Prefecture, Taiwan, ["Republic of China"]."

Phyllodactylus clinatus Dixon and Huey

Contrib. Sci., Nat. Hist. Mus. Los Angeles County 192:27. 1970.
HOLOTYPE: **MVZ 82227** (adult female); "Punta Aguja, 37 km SW [of] Sechura, Department of Piura, Peru"; R. B. Huey (343); July 19, 1967.

Phyllodactylus davisi Dixon

New Mexico State Univ. Res. Center Sci. Bull. 64-1:90. 1964.
PARATYPE: **MVZ 12186**; Jala, 21 km W of Colima, Colima, México, ca. 1,200 ft elevation.

Phyllodactylus xanti nocticolus Dixon

New Mexico State Univ. Res. Center Sci. Bull. 64-1:55. 1964.
CURRENT NAME: Murphy (1983:31) elevated this taxon to species level, *Phyllodactylus nocticolus* Dixon, but Grismer (1994) and Kluge (1993) referred to it as *Phyllodactylus xanti nocticolus* Dixon.
PARATYPES: **MVZ 61070-61072**; 10 mi W of The Narrows, San Diego County, California, USA.

Pseudogonatodes peruvianus Huey and Dixon

Copeia 1970:539. 1970.
HOLOTYPE: **MVZ 82136** (adult female); "Tingo, 5° 53′ S, 78° 12′ W, 1000 m [elevation], Rio Utcubamba; 30 km S, 41 km E [of] Bagua Grande, Department of Amazonas, Peru"; R. B. Huey (598); August 24, 1967.
PARATOPOTYPES: **MVZ 82137, 82139.**
REMARKS: MVZ 82137 is a cleared and stained skeletal preparation.

Iguanidae (sensu lato)

Crotaphytus collaris fuscus Ingram and Tanner

Brigham Young Univ. Sci. Bull., Biol. Ser. 13:23. 1971.
CURRENT NAME: *Crotaphytus collaris* Say (fide McGuire, 1996)
PARATYPE: **MVZ 70704**; 15 mi (by road) N of Chihuahua, Chihuahua, México.
REMARKS: MVZ 70704 was erroneously referred to as "UC 70704" in the original description, and the locality was given as "Chihuahua City."

Sceloporus hunsakeri Hall and Smith

Breviora 452:4. 1979.
HOLOTYPE: **MVZ 73570** (adult male); "3 mi. E of San Bartolo ± 500 ft. [elevation] [Baja California Sur, México]"; R. G. Crippen (277); February 17, 1960.
PARATOPOTYPES: **MVZ 73572, 73575, 73579-73584.**

Sceloporus occidentalis occidentalis Baird and Girard

[Bell.] Herpetologica 10:34. 1954.
NEOTYPE: **MVZ 59874** (adult male); "Benicia, [Solano County, California, USA]"; R. C. Stebbins (5692A); August 15, 1953.

Sceloporus occidentalis taylori Camp

Univ. California Publ. Zool. 17:66. 1916[c].
HOLOTYPE: **MVZ 5947** (adult male); "half way between Merced Lake and Sunrise Trail (Echo Creek basin), altitude 7500 feet, Yosemite National Park, [Mariposa County,] California[, USA]"; W. P. Taylor (7361); August 25, 1915.

Sceloporus scalaris brownorum Smith, Watkins-Colwell, Lemos-Espinal, and Chiszar

Southwest. Nat. 42:290. 1997.
PARATYPE: **MVZ 76550**; El Salto, Durango, México.

Stenocercus huancabambae Cadle

Proc. Acad. Nat. Sci. Philadelphia 143:30. 1991.
PARATYPES: **MVZ 82312-82315, 82369**; 8 km WSW of Bagua, Departamento Amazonas, Perú, 1,500 ft (457 m) elevation. **MVZ 82316-82317**; Río Chunchuca, 29 km S, 13 km W of Jaen, Departamento Cajamarca, Perú.

Stenocercus imitator Cadle

Proc. Acad. Nat. Sci. Philadelphia 143:39. 1991.
PARATYPES: **MVZ 82370, 89893**; 2 km W of Porculla Pass (30 km ENE of Olmos), Departamento Piura, Perú, 6,500 ft (1,980 m) elevation. **MVZ 89894-89897**; 35 mi WNW of Cajamarca, Departamento Cajamarca, Perú, 6,000 ft (1,830 m) elevation. As noted by Cadle (1991:39), field notes of C. B. Koford in the MVZ indicate that the locality of MVZ 89894-89897 lies in the Río Zaña valley, along the road between La Florida and Taulis, a hacienda of the upper Río Zaña, east of Monte Seco. The specimen tags of MVZ 82370 and 89893 list the locality as lying in "[Departamento] Lambayeque," but this is in error.

Stenocercus latebrosus Cadle

Bull. Mus. Comp. Zool. 155:268. 1998.
PARATYPES: **MVZ 119233-119236**; 5 mi SW of Otuzco, Departamento La Libertad, Perú, 8,000 ft (2,440 m) elevation (ca. 07° 55′ S, 78° 33′ W).

Stenocercus percultus Cadle

Proc. Acad. Nat. Sci. Philadelphia 143:18. 1991.
PARATYPES: **MVZ 82318-82320, 82365-82368, 119223-119231**; 2 km W of Porculla Pass (30 km ENE of Olmos), Departamento Piura, Perú, 6,500 ft (1,982 m) elevation. The specimen tags list this locality as lying in "[Departamento] Lambayeque," but this is in error, as noted by Cadle (1991:19).

Uma exsul Schmidt and Bogert

Amer. Mus. Novitates 1339:1. 1947.
PARATYPE: **MVZ 44332**; 13 mi NNE of San Pedro de las Colonias, Coahuila, México.
REMARKS: MVZ 44332 was formerly AMNH 67413.

Urosaurus lahtelai Rau and Loomis

J. Herpetol. 11:25. 1977.
PARATYPES: **MVZ 13362, 13366-13367, 13371, 13374-13375, 13377**; Cataviña, Baja California, México, 1,850 ft elevation. **MVZ 80380**; Cataviña, Baja California, México. **MVZ 116466**; 9.3 km S of turnoff to Rancho Santa Inés (=Ynez) on México Highway 1, SE of El Rosario, Baja California, México.

Uta ornata chiricahuae Mittleman

Proc. Biol. Soc. Washington 54:165. 1941.
CURRENT NAME: *Urosaurus ornatus linearis* (Baird)
HOLOTYPE: **MVZ 7751** (male); "Pinery Canyon, Chiricahua Mountains, 6,000 ft. [elevation], Cochise County, Arizona, [USA]"; J. E. Law (6963); May 10, 1919.
PARATOPOTYPES: **MVZ 7747-7750, 7752-7778, 8190.**
PARATYPES: **MVZ 13837-13840**; Dos Cabezas Mountains, Cochise County, Arizona, USA.

Uta stansburiana hesperis Richardson

Proc. U. S. Natl. Mus. 48:415. 1915.
CURRENT NAME: *Uta stansburiana elegans* Yarrow
HOLOTYPE: **MVZ 892** (male); "Arroyo Seco Canyon [=Arroyo Seco], near Pasadena, Los Angeles County, California, [USA]"; J. Grinnell; August 3, 1903.

Lacertidae

Platyplacopus kuehnei carinatus Gressitt

Proc. Biol. Soc. Washington 51:129. 1938[a].
HOLOTYPE: **MVZ 23519** (adult male); "Ta Han [=Da'an], alt. [=altitude] 775 meters, central part of Hainan Island [=Hainan Province], S. China [=South China Sea, China]"; J. L. Gressitt; June 23, 1935.

Scincidae

Emoia boettgeri orientalis Brown and Marshall

Copeia 1953:204. 1953.
PARATYPE: **MVZ 61158**; Ine Island, Arno Atoll, Marshall Islands.
REMARKS: The original description referred to several paratypes distributed among eleven collections but did not list them explicitly. MVZ 61158 matches the data presented by Brown and Marshall and therefore is believed to be a paratype.

Eumeces gilberti cancellosus Rodgers and Fitch

Univ. California Publ. Zool. 48:200. 1947.
HOLOTYPE: **MVZ 24034** (adult female); "8 mi. W and 1.1 mi. S [of] Altamont, 900 ft. [elevation], Alameda County, California[, USA]"; H. S. Fitch (3348); May 4, 1937. Our examination of maps from the area indicate that the type locality is 0.8 (not "8") mi W and 1.1 mi S of Altamont. The specimen tag and field notes of H. S. Fitch in the MVZ list the locality as "6 mi. E [of] Livermore," which is equivalent to 0.8 mi W and 1.1 mi S of Altamont.

Eumeces gilberti placerensis Rodgers

Copeia 1944:101. 1944.
HOLOTYPE: **MVZ 24058** (adult male); "6 miles east of Smartville, Nevada County, California, [USA]"; H. S. Fitch (3380); May 6, 1937.

Eumeces multivirgatus mexicanus Anderson and Wilhoft

Copeia 1959:57. 1959.
CURRENT NAME: *Eumeces multilineatus* Tanner
HOLOTYPE: **MVZ 66061** (adult male); "Yaguirachic, 29 degrees 35 minutes N. Lat. [=latitude], 108 degrees 10 minutes W. Long. [=longitude], 130 miles west [of] Chihuahua City, Chihuahua, Mexico"; J. D. Anderson (1566); July 2, 1957. The specimen tag lists the elevation as "8500 ft."
PARATOPOTYPES: **MVZ 66056-66060, 66062-66065**.

Fojia bumui Greer and Simon

J. Herpetol. 16:133. 1982.
PARATOPOTYPES: **MVZ 175806-175812**; "Moikisung area, Huon Peninsula, Morobe Province, Papua New Guinea. Elevation: 550 m. Coordinates: 147° 30′ E, 6° 34′ S."

Teiidae

Cnemidophorus neomexicanus Lowe and Zweifel

Bull. Chicago Acad. Sci. 9:230. 1952.
HOLOTYPE: **MVZ 55807** (adult female); "McDonald Ranch Headquarters, 4800 feet elevation, 8.7 miles west and 22.8 miles south of New Bingham Post Office, Socorro County, New Mexico, [USA]"; C. H. Lowe Jr. (3528); August 2, 1947. In 1945 the McDonald Ranch Headquarters were converted into the base camp for local operations of the U.S. Atomic Energy Commission.
PARATOPOTYPES: **MVZ 55788, 55790-55795, 55798-55804, 55806**.
REMARKS: The original description referred to several paratypes distributed among several collections but did not list them explicitly. MVZ 55788, 55790-55795, 55798-55804, and 55806 match the data for those discussed by Lowe and Zweifel and therefore are believed to be paratopotypes. MVZ 55807 was formerly UCLA 4588. MVZ 55790 and 55794 were sent on permanent loan to SU and now are CAS-SU 16104 and 16105, respectively.

Cnemidophorus sacki barrancorum Zweifel

Bull. Amer. Mus. Nat. Hist. 117:102. 1959.
CURRENT NAME: *Cnemidophorus costatus barrancorum* Zweifel
HOLOTYPE: **MVZ 50724** (adult male); "Rancho Guirocoba, about 20 miles southeast of Alamos, Sonora, Mexico"; R. G. Zweifel (1049); August 7, 1950.
PARATOPOTYPES: **MVZ 50725, 50727, 50734**.

Cnemidophorus sacki griseocephalus Zweifel

Bull. Amer. Mus. Nat. Hist. 117:96. 1959.
CURRENT NAME: *Cnemidophorus burti griseocephalus* Zweifel
PARATYPES: **MVZ 50719, 50721**; 12.4 mi NW of Navojoa, Sonora, México.

Cnemidophorus sacki mazatlanensis Zweifel

Bull. Amer. Mus. Nat. Hist. 117:89. 1959.
CURRENT NAME: *Cnemidophorus costatus mazatlanensis* Zweifel
PARATYPES: **MVZ 59183-59184**; 9 mi N of Mazatlán, Sinaloa, México, 500 ft elevation. **MVZ 59188, 59190-59195**; same locality as MVZ 59183-59184, 100 ft elevation.
REMARKS: MVZ 58194 (*Rana boylii*) was erroneously listed as a paratype in the original description. MVZ 59184 matches the data presented by Zweifel and therefore is believed to be a paratype. In addition, loan records in the MVZ show that Zweifel borrowed the latter specimen, not MVZ 58194.

Cnemidophorus sacki nigrigularis Zweifel

Bull. Amer. Mus. Nat. Hist. 117:93. 1959.
CURRENT NAME: *Cnemidophorus costatus nigrigularis* Zweifel
PARATYPES: **MVZ 59196-59197, 59203, 59205**; 25-26 mi NW of Elota, Sinaloa, México.
REMARKS: MVZ 59916 (*Hyla regilla*) was erroneously listed as a paratype in the original description. MVZ 59196 matches the data presented by Zweifel and therefore is believed to be a paratype. In addition, loan records in the MVZ show that Zweifel borrowed the latter specimen, not MVZ 59916.

Xantusiidae

Xantusia bezyi Papenfuss, Macey, and Schulte

Sci. Pap. Nat. Hist. Mus., Univ. Kansas 23:4. 2001.
HOLOTYPE: **MVZ 232604** (adult male); "33° 49.48′ N, 111° 28.55′ W, NE 1/4 Sec. 31, T. 6 N., R. 9 E., 5.6 km S (by Highway 87) of Sunflower, elev. [=elevation] 948 m, Maricopa County, Arizona, USA"; T. J. Papenfuss (26413); November 3, 2000.
PARATOPOTYPES: **MVZ 232605-232607**.
PARATYPES: **MVZ 232608-232611, 232571**; 2.9 km S (by Highway 87) of Sunflower, Maricopa County, Arizona, USA, 1,085 m elevation (NW 1/4 Sec. 29, T6N, R9E; 33° 51.10′ N, 111° 28.28′ W).

Xantusia riversiana Cope

Proc. Acad. Nat. Sci. Philadelphia 35:29. 1884 (1883).
HOLOTYPE: **MVZ 8278** (sex unknown); the type locality was given as "unknown, beyond that it is Californian." Later, Rivers (1889b:1100) restricted the type locality to "San Nicolas Island, the westward island of the Santa Barbara Group [=Channel Islands,] [Ventura County,] California[, USA]"; J. G. Cooper (T. I. Storer 1948); July 1863.
REMARKS: The name *Xantusia riversiana* was first used, without description, in an address by Edward D. Cope at the California Academy of Sciences in 1879 (Anonymous, 1879:801).

Xantusia vigilis wigginsi Savage

Amer. Midl. Nat. 48:473. 1952.
PARATYPES: **MVZ 58628-58629**; 18 mi S of Punta Prieta, Baja California Sur, México.
REMARKS: MVZ 58628-58629 were formerly SU 11556 and 11557, respectively.

SERPENTES

Colubridae

Chapinophis xanthocheilus Campbell and Smith

Herpetologica 54:210. 1998.
PARATYPE: **MVZ 160488**; 5 km ENE of Chilascó, Finca San Jorge, Departamento Baja Verapaz, Guatemala, 1,829 m elevation.

Chersodromus annulatus Zweifel

Herpetologica 10:17. 1954[b].
CURRENT NAME: *Sibon zweifeli* (Liner and Wilson)
HOLOTYPE: **MVZ 45030** (male); "near Chilpancingo, Guerrero, Mexico"; W. W. Brown; between August 1942 and August 1943.

Diadophis amabilis occidentalis Blanchard

Occ. Pap. Mus. Zool., Univ. Michigan 142:6. 1923.
CURRENT NAME: *Diadophis punctatus occidentalis* Blanchard
HOLOTYPE: **MVZ 7260** (female); "Bridgeville, Humboldt County, California[, USA]"; H. E. Wilder (1580); May 30, 1919.

Hypsiglena torquata tiburonensis Tanner

[W. Tanner.] Great Basin Nat. 41:139. 1981.
PARATYPE: **MVZ 37802**; Ensenada del Perro, south end of Isla Tiburón, Sonora, México.
REMARKS: MVZ 37802 was sent out on loan on March 7, 1967, and sent back to the MVZ on May 28, 1986, but the specimen was lost in the mail.

Lampropeltis getulus nigritus Zweifel and Norris

Amer. Midl. Nat. 54:238. 1955.
CURRENT NAME: *Lampropeltis getula nigrita* Zweifel and Norris
HOLOTYPE: **MVZ 50814** (adult male); "30.6 miles (by road) south of Hermosillo, Sonora, [México,] on the main highway"; K. S. Norris (1028) and R. G. Zweifel; August 3, 1950.

Lampropeltis zonata parvirubra Zweifel

Copeia 1952:160. 1952.
HOLOTYPE: **MVZ 42407** (adult male); "1/4 mi. NW of Falling Springs Resort, 2 mi. SW of Crystal Lake Park, San Gabriel Mountains, Los Angeles County, California[, USA]"; R. C. Stebbins (838); May 8, 1946.

Leptodeira ephippiata Smith and Tanner

Copeia 1944:131. 1944.
CURRENT NAME: *Leptodeira splendida ephippiata* Smith and Tanner
HOLOTYPE: **MVZ 28931** (female); "Agua Marín, 8.3 miles west northwest of Alamos, Sonora, [México]"; C. G. Sibley (426); May 5, 1939. The specimen tag lists the elevation as "1800' [feet] ±."

Liophis cobella dyticus Dixon

J. Herpetol. 17:159. 1983.
PARATYPE: **MVZ 123079**; 4 km NW of Pucallpa, Departamento Loreto, Perú, 450 ft elevation.

Masticophis lateralis euryxanthus Riemer

Copeia 1954:45. 1954.
HOLOTYPE: **MVZ 50391** (adult female); "Berkeley Hills, Berkeley, Alameda County, California[, USA]"; A. Mossman (W. J. Riemer 893); August 15, 1950.

Natrix boulengeri Gressitt

Proc. Biol. Soc. Washington 50:125. 1937.
CURRENT NAME: *Amphiesma boulengeri* (Gressitt)
HOLOTYPE: **MVZ 23623** (adult female); "Tai-yong [=Dayang, formerly Jieyang County, now Jiexi County], alt. [=altitude] 640 meters, E. [=eastern] Kwangtung province [=Guangdong Province], southeastern China (lat. [=latitude] 23° 34' N., long. [=longitude] 115° 55' E.)"; J. L. Gressitt (1487); August 5, 1936.
PARATYPE: **MVZ 23622**; Hong San (=Xunwu County), Kiangsi (=Jiangxi) Province, China, 850 m elevation (24° 58' N, 115° 50' E).

Pliocercus elapoides celatus Smith

J. Washington Acad. Sci. 33:344. 1943.
CURRENT NAME: *Pliocercus bicolor bicolor* Smith (fide Smith and Chiszar, 1996; see also Smith and Chiszar, 2001); *Pliocercus elapoides* Cope of some authors. Savage and Crother (1989) placed *Pliocercus* in the synonymy of *Urotheca* Bibron, but other workers (e.g., C. Myers and Cadle, 1994; Smith et al., 1995) argued in favor of retaining *Pliocercus*.
HOLOTYPE: **MVZ 24689** (male); "Ciudad Victoria, Tamaulipas, Mexico"; M. Embury; June 31, 1937. The specimen tag lists the collection date as June 21, 1937, but we were unable to determine which date is correct.

Psammodynastes pulverulentus papenfussi Zhao

Sichuan J. Zool. 14:108. 1995.
HOLOTYPE: **MVZ 23857** (adult female); "Kuraru (=Kueitzuchiao), Koshun District (=Henchun Town), Takao-shu Province (=Pintung Hsien), Taiwan Province [=Taiwan, "Republic of China"], 150 meters [elevation]"; J. L. Gressitt (63); May 5, 1934.

Rhinocheilus lecontei clarus Klauber

Trans. San Diego Soc. Nat. Hist. 9:308. 1941.
CURRENT NAME: *Rhinocheilus lecontei lecontei* Baird and Girard
PARATYPE: **MVZ 228**; Dos Palmas Spring, Santa Rosa Mountains, Riverside County, California, USA.

Sonora mosaueri Stickel

Copeia 1938:189. 1938.
CURRENT NAME: *Sonora semiannulata* Baird and Girard
HOLOTYPE: **MVZ 13772** (male); "[San Jose de] Comondu, Lower [=Baja] California [Sur,] [México]"; C. C. Lamb (14164); April 9, 1931. The specimen tag lists the elevation as "1000 ft." The original description lists the collection date as April 2, 1931, but the specimen tag and field notes of C. C. Lamb in the MVZ indicate that the correct collection date is April 9, 1931.
PARATOPOTYPES: **MVZ 13770-13771, 13773**.

Sonora semiannulata gloydi Stickel

Copeia 1938:186. 1938.
CURRENT NAME: *Sonora semiannulata* Baird and Girard
PARATYPE: **MVZ 17580**; Bright Angel Trail, Grand Canyon National Park, Coconino County, Arizona, USA.

Tantilla cuesta Wilson

Contrib. Biol. Geol., Milwaukee Public Mus. 52:29. 1982.
HOLOTYPE: **MVZ 146762** (adult? female); "Finca Santa Julia, 1.5 km E [of] San Rafael Pie de la Cuesta, Depto. [=Departamento] San Marcos, Guatemala, elevation 1050 m"; J. E. Cadle (1562); August 27, 1977.
PARATOPOTYPE: **MVZ 146763**.

Tantilla excubitor Wilson

Contrib. Biol. Geol., Milwaukee Public Mus. 52:37. 1982.
CURRENT NAME: *Tantillita brevissima* (Taylor)
HOLOTYPE: **MVZ 88468** (adult? male); "Finca El Salto, 2 km E [of] Escuintla, Depto. [=Departamento] Escuintla, Guatemala, elevation 305 m"; J. E. Woods (G-110); mid-March 1969.

Tantilla tayrae Wilson

J. Herpetol. 17:54. 1983.
HOLOTYPE: **MVZ 159203** (adult male); "Finca San Jerónimo, 7.5 km N (by road) [of] Cacaohatán (=Cacahoatán or Cacahuatán), elevation 760 m, Volcán Tacaná, Municipio de Unión Juárez, Chiapas, México"; R. L. Seib (3537); July 30, 1978.
PARATOPOTYPES: **MVZ 159114, 167117, 169587-169588**.
REMARKS: The original description lists the collection date of MVZ 169587 as "14 June 1972," but the specimen tag and field notes of R. L. Seib in the MVZ clearly indicate that the correct collection date is June 14, 1979.

Thamnophis atratus zaxanthus Boundy

Proc. California Acad. Sci. 51:328. 1999.
HOLOTYPE: **MVZ 207940** (adult female); "3 mi[les] S of Gilroy Hot Springs, Santa Clara County, California, [USA]"; W. P. Hutchins (3086); April 15, 1986. The specimen tag lists the collector as "B. Hutchins," instead of "W. P. Hutchins."

Thamnophis elegans aquaticus Fox

Univ. California Publ. Zool. 50:493. 1951[a].
CURRENT NAME: *Thamnophis   a.   atratus* (Kennicott) x *T. a. hydrophilus* (Fitch) intergrades
HOLOTYPE: **MVZ 48196** (adult male); "Dillon Beach, Marin County, California[, USA]"; W. Fox (2753); July 18, 1949.

Thamnophis elegans nigrescens Johnson

[M. Johnson.] Herpetologica 3:161. 1947.
CURRENT NAME: *Thamnophis elegans vagrans* (Baird and Girard)
PARATOPOTYPE: **MVZ 44658**; "Tacoma, [Pierce County,] Washington[, USA]."
REMARKS: MVZ 44658 was formerly CPS 3113.

Thamnophis elegans terrestris Fox

Univ. California Publ. Zool. 50:499. 1951[a].
HOLOTYPE: **MVZ 48197** (adult female); "Strawberry Canyon, Berkeley, Alameda County, California[, USA]"; W. Fox (2744); June 23, 1949.

Thamnophis ordinoides gigas Fitch

Univ. California Publ. Zool. 44:69. 1940.
CURRENT NAME: *Thamnophis gigas* Fitch
HOLOTYPE: **MVZ 5428** (adult female); "Gadwall, Merced County, California[, USA]"; H. C. Bryant (688); May 16, 1914.

Thamnophis ordinoides hydrophila Fitch

Amer. Midl. Nat. 17:648. 1936.
CURRENT NAME: *Thamnophis atratus hydrophilus* Fitch
HOLOTYPE: **MVZ 18127** (adult male); "Trail Creek 6 miles from its mouth, Jackson County, Oregon[, USA]"; H. S. Fitch (2361); July 27, 1934.

Thamnophis sirtalis fitchi Fox

Copeia 1951:264. 1951[b].
HOLOTYPE: **MVZ 51778** (adult female); "Greylodge Refuge, 9 mi. W of Gridley, Butte Co., California[, USA]"; J. Cowan (W. Fox 2953); June 17, 1950.

Elapidae

Micruroides euryxanthus australis Zweifel and Norris

Amer. Midl. Nat. 54:246. 1955.
HOLOTYPE: **MVZ 50839** (adult male); "purchase[d] from a native collector at Guirocoba, Sonora, Mexico"; R. G. Zweifel (1110) and K. S. Norris; August 10, 1950.
PARATOPOTYPES: **MVZ 50838, 50840.**

Leptotyphlopidae

Leptotyphlops humilis chihuahuaensis Tanner

[W. Tanner.] Great Basin Nat. 45:623. 1985.
PARATYPE: **MVZ 57331**; 3 mi NW of Chilmahma, a northwestern suburb of Chihuahua City, 0.5 mi W of main highway, Chihuahua, México.

Typhlopidae

Typhlops adamsi Tanner

[V. Tanner.] Great Basin Nat. 11:64. 1951.
CURRENT NAME: *Acutotyphlops infralabialis* (Waite)
HOLOTYPE: **MVZ 40753** (female); "Guadalcanal[,] Nalimbiu River, Solomon Islands"; L. Adams (6); June 6, 1944. The specimen tag lists the locality as "1 mi. Inland" and the elevation as "50 ft."

Viperidae

Bothrops tzotzilorum Campbell

J. Herpetol. 19:48. 1985.
CURRENT NAME: *Cerrophidion tzotzilorum* (Campbell)
PARATYPE: **MVZ 57264**; 9.7 km SE of San Cristóbal de las Casas, Chiapas, México.

Crotalus cerastes cercobombus Savage and Cliff

Nat. Hist. Misc. 119:2. 1953.
PARATYPE: **MVZ 58630**; 2 mi E of Gunsight, Pima County, Arizona, USA.
REMARKS: MVZ 58630 was formerly SU 13998.

Crotalus confluentus stephensi Klauber

Trans. San Diego Soc. Nat. Hist. 6:108. 1930.
CURRENT NAME: *Crotalus mitchellii stephensi* Klauber
HOLOTYPE: **MVZ 6699** (adult male); "two miles west of Jackass Springs, Panamint Mts. [=Mountains], altitude 6200 ft., Inyo County, California, [USA]"; J. Grinnell (4597); October 8, 1917.

Crotalus enyo furvus Lowe and Norris

Trans. San Diego Soc. Nat. Hist. 12:52. 1954.
CURRENT NAME: Beaman and Grismer (1994:589.4) stated that *C. enyo enyo* and *C. enyo furvus* "should be considered as the binomial *C. enyo*," but McDiarmid et al. (1999:284) retained the two subspecies pending publication of supportive data.
HOLOTYPE: **MVZ 55388** (adult male); "10.9 miles (by road) north of El Rosario, along the main road on the coastwise terrace near the foot of a bold Cretaceous escarpment, Baja California Norte [=Baja California], Mexico"; K. S. Norris (786) and C. H. Lowe Jr.; July 21, 1949.

Crotalus viridis caliginis Klauber

Trans. San Diego Soc. Nat. Hist. 11:90. 1949[a].
PARATOPOTYPE: **MVZ 5404**; "South Coronado Island [=South Island, Los Coronados Islands], off the northwest coast of Baja California, Mexico."

Crotalus willardi amabilis Anderson

Copeia 1962:160. 1962.
HOLOTYPE: **MVZ 68896** (adult male); "Arroyo Mesteño, 8,500 feet [elevation], Sierra del Nido, Chihuahua, Mexico"; G. A. Bryan (91); July 12, 1960. The specimen tag lists the elevation as 7,600 ft and the collection year as 1959, but we were unable to determine which information is correct.
PARATOPOTYPES: **MVZ 68895, 68897-68900, 71015-71016.**
PARATYPES: **MVZ 66117, 68894**; Cañón del Álamo, west side of the Sierra del Nido, Chihuahua, México, 8,000 ft elevation. **MVZ 68893**; Cañón

del Álamo, west side of the Sierra del Nido, Chihuahua, México, 7,300 ft elevation.

Crotalus willardi silus Klauber

Trans. San Diego Soc. Nat. Hist. 11:128. 1949[b].
HOLOTYPE: **MVZ 46694** (adult male); "Río Gavilán, 7 miles southwest of Pacheco, Chihuahua, Mexico, altitude 6200 ft."; R. McCabe (W. C. Russell 10990); August 13, 1948.
PARATYPES: **MVZ 46692, 46695-46696**; type locality, 6,700 ft elevation. **MVZ 46693**; type locality, 5,700 ft elevation.

TESTUDINES

Chelidae

Platemys platycephala melanonota Ernst

J. Herpetol. 17:352. 1984 (1983).
PARATYPES: **MVZ 158995**; east bank of Río Santiago at La Poza, Río Santiago drainage, Departamento Amazonas, Perú. **MVZ 163036, 163039**; vicinity of San Antonio, Río Cenepa, Río Cenepa drainage, Departamento Amazonas, Perú. **MVZ 163037**; Río Huampami (tributary of Río Cenepa), Río Cenepa drainage, Departamento Amazonas, Perú. **MVZ 163038**; vicinity of Huampami, Río Cenepa, Río Cenepa drainage, Departamento Amazonas, Perú. **MVZ 163040**; Tujushik Entsa, near Huampami, Río Cenepa, Río Cenepa drainage, Departamento Amazonas, Perú. **MVZ 163041**; vicinity of Kusu, Río Comaina, Río Cenepa drainage, Departamento Amazonas, Perú. **MVZ 163042**; vicinity of Huampami, Río Cenepa, Río Cenepa drainage, Departamento Amazonas, Perú. **MVZ 175379**; vicinity of La Poza, Río Santiago, Río Santiago drainage, Departamento Amazonas, Perú.

Emydidae

Clemmys marmorata pallida Seeliger

Copeia 1945:158. 1945.
HOLOTYPE: **MVZ 6716** (adult female); "Lower Coyote Creek, near Alamitos, Orange County, California, [USA]"; J. E. Law; summer 1916 or 1917.

Trionychidae

Aspidonectes californiana Rivers

Proc. California Acad. Sci. Ser. 2, 2:233. 1889[a].

CURRENT NAME: *Palea steindachneri* (Siebenrock)

REMARKS: Rivers described *Aspidonectes californiana* based on a single specimen collected in the Sacramento River, northern California. Rivers (1889a:235) stated that the collectors sent the yet uncataloged specimen to the Museum of the University of California (the precursor of the MVZ), but there is no evidence that the specimen was transferred from the original museum to the MVZ, as were some other specimens. We were unable to locate this specimen, which is presumed lost (Webb, 1975).

GEOGRAPHIC DISTRIBUTION OF TYPE LOCALITIES OF RECENT AMPHIBIANS AND NONAVIAN REPTILES IN THE MUSEUM OF VERTEBRATE ZOOLOGY

The following list includes all taxa known to be present in the Museum of Vertebrate Zoology as of December 31, 2001, whose geographic place of capture constitutes a type locality. The nature of the type material in the MVZ—holotype, neotype, syntypes, or paratopotype(s)—is indicated by an "H," "N," "S," or "P," respectively, in parentheses after the name of each taxon as given in the original description.

China

Microhylidae
Kaloula pulchra hainana Gressitt (H, P)

Hynobiidae
Batrachuperus taibaiensis Song, Zeng, Wu, Liu, and Fu (P)

Salamandridae
Cynops orphicus Risch (H, P)

Lacertidae
Platyplacopus kuehnei carinatus Gressitt (H)

Colubridae
Natrix boulengeri Gressitt (H)
[Current name: *Amphiesma boulengeri* (Gressitt)]

Costa Rica

Plethodontidae
Bolitoglossa gracilis Bolaños, Robinson, and Wake (P)
Nototriton gamezi García-París and Wake (H)
Nototriton guanacaste Good and Wake (H, P)
Nototriton tapanti Good and Wake (H)

Guatemala

Hylidae
Hyla minera Wilson, McCranie, and Williams (H)

Plethodontidae
Bolitoglossa jacksoni Elias (H)
Bolitoglossa meliana Wake and Lynch (H, P)
Bradytriton silus Wake and Elias (H, P)
Chiropterotriton cuchumatanus Lynch and Wake (H, P)
[Current name: *Dendrotriton cuchumatanus* (Lynch and Wake)]
Chiropterotriton rabbi Lynch and Wake (H, P)
[Current name: *Dendrotriton rabbi* (Lynch and Wake)]
Chiropterotriton veraepacis Lynch and Wake (H, P)
[Current name: *Cryptotriton veraepacis* (Lynch and Wake)]
Nyctanolis pernix Elias and Wake (H, P)
Spelerpes dofleini Werner (N)
[Current name: *Bolitoglossa dofleini* (Werner)]

Colubridae
Tantilla cuesta Wilson (H, P)
Tantilla excubitor Wilson (H)
[Current name: *Tantillita brevissima* (Taylor)]

Honduras

Leptodactylidae
Eleutherodactylus omoaensis McCranie and Wilson (H, P)

Plethodontidae
Bolitoglossa diaphora McCranie and Wilson (H, P)
Bolitoglossa longissima McCranie and Cruz (H)

Iran

Salamandridae
Rhithrotriton derjugini microspilotus Nesterov (S)
[Current name: *Neurergus microspilotus* (Nesterov)]

México

Hylidae
Hyla californiae Gorman (H)
[Current name: *Hyla cadaverina* Cope; *Pseudacris cadaverina* (Cope) of some authors]

Ranidae
Rana pueblae Zweifel (P)
Rana sinaloae Zweifel (H, P)
[Current name: *Rana pustulosa* Boulenger]
Rana tlaloci Hillis and Frost (P)

Plethodontidae
Bolitoglossa hermosa Papenfuss, Wake, and Adler (H)
Ixalotriton niger Wake and Johnson (H, P)
Nototriton adelos Papenfuss and Wake (H, P)
[Current name: *Cryptotriton adelos* (Papenfuss and Wake)]
Nototriton alvarezdeltoroi Papenfuss and Wake (H)
[Current name: *Cryptotriton alvarezdeltoroi* (Papenfuss and Wake)]
Pseudoeurycea aquatica Wake and Campbell (P)
Pseudoeurycea longicauda Lynch, Wake, and Yang (H, P)
Pseudoeurycea lynchi Parra-Olea, Papenfuss, and Wake (H, P)
Pseudoeurycea naucampatepetl Parra-Olea, Papenfuss, and Wake (H, P)
Pseudoeurycea parva Lynch and Wake (H, P)
[Current name: *Ixalotriton parvus* (Lynch and Wake)]
Pseudoeurycea saltator Lynch and Wake (H)
Thorius arboreus Hanken and Wake (H)
Thorius aureus Hanken and Wake (H, P)
Thorius boreas Hanken and Wake (H, P)
Thorius grandis Hanken, Wake, and Freeman (H, P)
Thorius infernalis Hanken, Wake, and Freeman (H, P)
Thorius lunaris Hanken and Wake (H, P)
Thorius magnipes Hanken and Wake (H, P)
Thorius minydemus Hanken and Wake (H)
Thorius munificus Hanken and Wake (H, P)
Thorius omiltemi Hanken, Wake, and Freeman (H, P)
Thorius papaloae Hanken and Wake (H, P)
Thorius smithi Hanken and Wake (H)
Thorius spilogaster Hanken and Wake (H, P)

México (continued)

Anguidae
Abronia kalaina Good and Schwenk (H)
[Current name: *Abronia fuscolabialis* (Tihen)]
Gerrhonotus paucicarinatus Fitch (H)
[Current name: *Elgaria paucicarinata* (Fitch)]
Gerrhonotus scincicauda nanus Fitch (H)
[Current name: *Elgaria multicarinata nana* (Fitch)]

Iguanidae (sensu lato)
Sceloporus hunsakeri Hall and Smith (H, P)

Scincidae
Eumeces multivirgatus mexicanus Anderson and Wilhoft (H, P)
[Current name: *Eumeces multilineatus* Tanner]

Teiidae
Cnemidophorus sacki barrancorum Zweifel (H, P)
[Current name: *Cnemidophorus costatus barrancorum* Zweifel]

Colubridae
Chersodromus annulatus Zweifel (H)
[Current name: *Sibon zweifeli* (Liner and Wilson)]
Lampropeltis getulus nigritus Zweifel and Norris (H)
[Current name: *Lampropeltis getula nigrita* Zweifel and Norris]
Leptodeira ephippiata Smith and Tanner (H)
[Current name: *Leptodeira splendida ephippiata* Smith and Tanner]
Pliocercus elapoides celatus Smith (H)
[Current name: *Pliocercus bicolor bicolor* Smith (fide Smith and Chiszar, 1996); *Pliocercus elapoides* Cope of some authors]
Sonora mosaueri Stickel (H, P)
[Current name: *Sonora semiannulata* Baird and Girard]
Tantilla tayrae Wilson (H, P)

Elapidae
Micruroides euryxanthus australis Zweifel and Norris (H, P)

Viperidae
Crotalus enyo furvus Lowe and Norris (H)
Crotalus viridis caliginis Klauber (P)
Crotalus willardi amabilis Anderson (H, P)
Crotalus willardi silus Klauber (H)

Morocco

Amphisbaenidae
Blanus tingitanus Busack (H, P)

Panamá

Plethodontidae
Oedipina maritima García-París and Wake (P)

Papua New Guinea

Scincidae
Fojia bumui Greer and Simon (P)

Perú

Plethodontidae
Bolitoglossa digitigrada Wake, Brame, and Thomas (P)

Gekkonidae
Phyllodactylus clinatus Dixon and Huey (H)
Pseudogonatodes peruvianus Huey and Dixon (H, P)

Solomon Islands

Typhlopidae
Typhlops adamsi Tanner (H)
[Current name: *Acutotyphlops infralabialis* (Waite)]

Taiwan, "Republic of China"

Microhylidae
Rana gracilipes Gressitt (H)
[Current name: *Micryletta steinegeri* (Boulenger)]

Agamidae
Japalura brevipes Gressitt (P)

Gekkonidae
Hemidactylus stejnegeri Ota and Hikida (P)

Taiwan, "Republic of China" (continued)

Colubridae
  Psammodynastes pulverulentus papenfussi Zhao (H)

Tanzania

Scolecomorphidae
  Scolecomorphus uluguruensis Barbour and Loveridge (P)

United States

Arizona

  Iguanidae (sensu lato)
    Uta ornata chiricahuae Mittleman (H, P)
    [Current name: *Urosaurus ornatus linearis* (Baird)]

  Xantusiidae
    Xantusia bezyi Papenfuss, Macey, and Schulte (H, P)

California

  Ascaphidae
    Ascaphus truei californicus Mittleman and Myers (H)
    [Current name: *Ascaphus truei* Stejneger]

  Bufonidae
    Bufo canorus Camp (H)
    Bufo cognatus californicus Camp (H)
    [Current name: *Bufo californicus* Camp]
    Bufo exsul Myers (P)

  Ranidae
    Rana boylii muscosa Camp (H)
    [Current name: *Rana muscosa* Camp]
    Rana boylii sierrae Camp (H)
    [Current name: *Rana muscosa* Camp]

  Ambystomatidae
    Ambystoma macrodactylum croceum Russell and Anderson (H, P)

United States (continued)

California (continued)

Plethodontidae
Aneides vagrans Wake and Jackman (H, P)
Batrachoseps campi Marlow, Brode, and Wake (H, P)
Batrachoseps diabolicus Jockusch, Wake, and Yanev (H, P)
Batrachoseps gabrieli Wake (H, P)
Batrachoseps gavilanensis Jockusch, Yanev, and Wake (H, P)
Batrachoseps gregarius Jockusch, Wake, and Yanev (H, P)
Batrachoseps incognitus Jockusch, Yanev, and Wake (H, P)
Batrachoseps kawia Jockusch, Wake, and Yanev (H, P)
Batrachoseps luciae Jockusch, Yanev, and Wake (H, P)
Batrachoseps major Camp (H)
Batrachoseps minor Jockusch, Yanev, and Wake (H, P)
Batrachoseps regius Jockusch, Wake, and Yanev (H, P)
Batrachoseps stebbinsi Brame and Murray (H, P)
Ensatina eschscholtzii picta Wood (H)
Ensatina eschscholtzii xanthoptica Stebbins (H)
Ensatina sierrae Storer (H)
[Current name: *Ensatina eschscholtzii platensis* (Espada)]
Hydromantes brunus Gorman (H)
Hydromantes shastae Gorman and Camp (H, P)
Spelerpes platycephalus Camp (H, P)
[Current name: *Hydromantes platycephalus* (Camp)]

Rhyacotritonidae
Rhyacotriton olympicus variegatus Stebbins and Lowe (H)
[Current name: *Rhyacotriton variegatus* Stebbins and Lowe]

Salamandridae
Triturus rivularis Twitty (H, P)
[Current name: *Taricha rivularis* (Twitty)]
Triturus similans Twitty (H, P)
[Current name: *Taricha granulosa granulosa* (Skilton)]

Anguidae
Anniella nigra argentea Hunt (H)
[Current name: *Anniella pulchra* Gray]
Anniella pulchra Gray (N)
Gerrhonotus coeruleus shastensis Fitch (H)
[Current name: *Elgaria coerulea shastensis* (Fitch)]
Gerrhonotus panamintinus Stebbins (H, P)
[Current name: *Elgaria panamintina* (Stebbins)]

United States (continued)

California (continued)

Iguanidae (sensu lato)
Sceloporus occidentalis occidentalis Baird and Girard (N)
Sceloporus occidentalis taylori Camp (H)
Uta stansburiana hesperis Richardson (H)
[Current name: *Uta stansburiana elegans* Yarrow]

Scincidae
Eumeces gilberti cancellosus Rodgers and Fitch (H)
Eumeces gilberti placerensis Rodgers (H)

Xantusiidae
Xantusia riversiana Cope (H)

Colubridae
Diadophis amabilis occidentalis Blanchard (H)
[Current name: *Diadophis punctatus occidentalis* Blanchard]
Lampropeltis zonata parvirubra Zweifel (H)
Masticophis lateralis euryxanthus Riemer (H)
Thamnophis atratus zaxanthus Boundy (H)
Thamnophis elegans aquaticus Fox (H)
[Current name: *Thamnophis a. atratus* (Kennicott) x *T. a. hydrophilus* (Fitch) intergrades]
Thamnophis elegans terrestris Fox (H)
Thamnophis ordinoides gigas Fitch (H)
[Current name: *Thamnophis gigas* Fitch]
Thamnophis sirtalis fitchi Fox (H)

Viperidae
Crotalus confluentus stephensi Klauber (H)
[Current name: *Crotalus mitchellii stephensi* Klauber]

Emydidae
Clemmys marmorata pallida Seeliger (H)

New Mexico

Plethodontidae
Plethodon neomexicanus Stebbins and Riemer (H, P)

Teiidae
Cnemidophorus neomexicanus Lowe and Zweifel (H, P)

United States (continued)

North Carolina

Plethodontidae
Plethodon longicrus Adler and Dennis (P)
[Current name: *Plethodon yonahlossee* Dunn]

Oregon

Rhyacotritonidae
Rhyacotriton cascadae Good and Wake (H, P)
Rhyacotriton kezeri Good and Wake (H, P)

Colubridae
Thamnophis ordinoides hydrophila Fitch (H)
[Current name: *Thamnophis atratus hydrophilus* Fitch]

Texas

Plethodontidae
Eurycea tonkawae Chippindale, Price, Wiens, and Hillis (P)

Utah

Ambystomatidae
Ambystoma tigrinum utahense Lowe (H)
[Current name: *Ambystoma tigrinum nebulosum* Hallowell]

Virginia

Plethodontidae
Plethodon richmondi popei Highton and Grobman (P)

Washington

Colubridae
Thamnophis elegans nigrescens Johnson (P)
[Current name: *Thamnophis elegans vagrans* (Baird and Girard)]

Vietnam

Megophryidae
Leptolalax sungi Lathrop, Murphy, Orlov, and Ho (P)

Ranidae
Rana attigua Inger, Orlov, and Darevsky (P)

Vietnam (continued)

Rhacophoridae
  Philautus abditus Inger, Orlov, and Darevsky (P)
  Rhacophorus baliogaster Inger, Orlov, and Darevsky (P)

APPENDIX – HERPETOLOGISTS IN THE MUSEUM OF VERTEBRATE ZOOLOGY

The following is a list of curators, graduate and undergraduate students, postdoctoral fellows, research associates, research assistants, curatorial associates, curatorial assistants, and visiting faculty who have conducted research on the biology of amphibians and reptiles while in residence in the Museum of Vertebrate Zoology as of December 31, 2001, even if most of their work has involved other animal groups. In the case of former and current undergraduate and graduate students, the name(s) in parentheses following the dates of tenure is (are) the faculty member(s) under whose direction the student worked. Asterisks indicate that additional information about that person is provided in the section *Brief History of Herpetology in the Museum of Vertebrate Zoology*.

Adams, Lowell W.
Master's student, ?–1939 (E. Raymond Hall).
M.A. thesis: The Natural History and Classification of the Mount Lyell Salamander, *Hydromantes platycephalus* (Camp).

Alberch, Pedro (Pere)
Ph.D. student, 1976–1980 (George F. Oster and David B. Wake); Curatorial Assistant of Herpetology, 1976–1977.
Ph.D. dissertation: Heterochrony and Adaptation in the Evolution of *Bolitoglossa* (Amphibia: Caudata).

Alexandrino, João M.
Fundação para a Ciência e Tecnologia (Portugal) Postdoctoral Fellow, 2001–present.

77

Anderson, James D.
Ph.D. student, 1954–1960 (Robert C. Stebbins); Curatorial Assistant of Herpetology 1958–1959.
Ph.D. dissertation: A Comparative Study of Coastal and Montane Populations of *Ambystoma macrodactylum*.

Anderson, Paul K.
Ph.D. student, 1952–1958 (Robert C. Stebbins); Curatorial Assistant of Herpetology, 1956.
Ph.D. dissertation: Ecology and Evolution in Island Populations of Salamanders.

Arnold, Stevan J.
Undergraduate Curatorial Assistant of Herpetology, 1962–1966; Miller Postdoctoral Fellow, 1971–1973.

Atsatt, Sarah R.*
Master's student, 1910–1912 (Charles A. Kofoid); Ph.D. student, 1922–1931 (Samuel J. Holmes).
M.S. thesis: The Lacertilia of the San Jacinto Region.
Ph.D. dissertation: Color Changes as Controlled by Temperature and Light in the Lizards of the Desert Regions of Southern California.

Autumn, Kellar
Ph.D. student, 1989–1995 (Robert J. Full and Harry W. Greene); Curatorial Assistant of Herpetology, 1990; Instructional Technology Program (UC Berkeley) Postdoctoral Fellow, 1996.
Ph.D. dissertation: Performance at Low Temperatures and the Evolution of Nocturnality in Lizards.

Balgooyen, Thomas G.
Ph.D. student, 1969–1972 (A. Starker Leopold).
Ph.D. dissertation: Behavior and Ecology of the American Kestrel (*Falco sparverius*).

Barber, Paul H.
Ph.D. student, 1991–1996 (Anthony D. Barnosky); 1996–1998 (Tyrone B. Hayes).
Ph.D. dissertation: Phylogeography, Gene Flow, and Evolutionary History of the Canyon Treefrog, *Hyla arenicolor* (Cope).

Barlow, George W.
Research Ethologist, 1967–present.

Barwick, Richard E.
Visiting Research Associate, 1965.

Bauer, Aaron M.
Ph.D. student, 1982–1986 (Marvalee H. Wake).
Ph.D. dissertation: Systematics, Biogeography and Evolutionary Morphology of the Carphodactylini (Reptilia: Gekkonidae).

Bell, Christopher J.
Ph.D. student, 1992–1997 (Anthony D. Barnosky).
Ph.D. dissertation: A Revision of North American Irvingtonian (Early and Middle Pleistocene) Microtine Rodent Biochronology.

Bello, Roberto E.
Master's student, 1996–2001 (David B. Wake).
M.A. thesis: Prey Capture in Bolitoglossine Salamanders: Kinematics and Feeding Behavior.

Bemis, William E.
Ph.D. student, 1979–1982 (Marvalee H. Wake).
Ph.D. dissertation: Studies on the Evolutionary Morphology of Lepidosirenid Lungfish (Pisces: Dipnoi).

Bennett, Albert F.
Miller Postdoctoral Fellow, 1971–1973.

Berry, Kristin H.
Ph.D. student, 1968–1972 (Robert C. Stebbins).
Ph.D. dissertation: The Ecology and Social Behavior of the Chuckwalla, *Sauromalus obesus*.

Bingham, Robert E.
Ph.D. student, 2001–present (Craig Moritz and David B. Wake).
Ongoing dissertation research topic: Population dynamics and persistence of *Sceloporus* lizards in fragmented habitats in California.

Brodie, Edmund D., III
Miller Postdoctoral Fellow, 1991–1993.

Brown, Allen G.
Ph.D. student, 1961–1972 (Robert C. Stebbins).
Ph.D. dissertation: Responses to Problems of Water and Electrolyte Balance by Salamanders (Genus *Aneides*) from Different Habitats.

Brown, Charles W.
Ph.D. student, 1961–1970 (Robert C. Stebbins); Curatorial Assistant of Herpetology, 1961–1962, 1965, 1967–1968.
Ph.D. dissertation: Hybridization among the Subspecies of the Plethodontid Salamander *Ensatina eschscholtzii*.

Bryant, Harold C.
Master's student, 1909–1910 (Charles A. Kofoid).
M.S. thesis: The Horned Lizards (*Phrynosoma, Cleisope*) of California and Nevada.

Buchholz, Daniel R.
Ph.D. student, 1995–1999 (Tyrone B. Hayes).
Ph.D. dissertation: Evolution and Endocrinology of Accelerated Metamorphosis in Spadefoot Toads (Anura: Pelobatidae).

Bunnell, Pille L.
Ph.D. student, 1966–1973 (Oliver P. Pearson).
Ph.D. dissertation: A Stochastic Model of the Behavior of a Lizard Community.

Bury, R. Bruce
Ph.D. student, 1967–1972 (Robert C. Stebbins); Curatorial Assistant of Herpetology, 1968–1969, 1971–1972.
Ph.D. dissertation: Habits and Home Range of the Pacific Pond Turtle, *Clemmys marmorata*, in a Stream Community.

Busack, Stephen D.*
Ph.D. student, 1978–1985 (David B. Wake); Curatorial Assistant of Herpetology, 1978–1984.
Ph.D. dissertation: A Biogeographical Analysis of a Vicariant Event: The Herpetofauna of the Gibraltar Strait.

Cadle, John E.*
Ph.D. student, 1976–1982 (David B. Wake); Curatorial Assistant of Herpetology, 1976–1978, 1980–1982; Research Fellow, 1983–1984.
Ph.D. dissertation: Evolutionary Relationships among Advanced Snakes.

Camp, Charles L.*
Undergraduate Curatorial Assistant of Herpetology, 1911, 1915–1916.

Cannatella, David C.
National Science Foundation Postdoctoral Fellow, 1986–1988.

Carothers, John H.
Ph.D. student, 1980–1987 (Harry W. Greene); Curatorial Assistant of Herpetology, 1980–1984, 1986.
Ph.D. dissertation: Aspects of the Ecology of Lizards of the Genus *Liolaemus* in the Central Chilean Cordillera.

Carrier, David
National Institutes of Health Postdoctoral Fellow, 1988–1991.

Case, Susan M.
Ph.D. student, 1971–1976 (James L. Patton).
Ph.D. dissertation: Evolutionary Studies in Selected North American Frogs of the Genus *Rana* (Amphibia, Anura).

Charland, M. Brent
Natural Sciences and Engineering Research Council of Canada Postdoctoral Fellow, 1992–1994.

Cohen, Nathan W.
Master's student, 1947–1950 (Robert C. Stebbins); Curatorial Assistant of Herpetology, 1949; Research Associate of Herpetology, 1970–1980.
M.A. thesis: Comparative Rates of Dehydration and Hydration in Some California Salamanders.

Collazo, Andrés
Ph.D. student, 1985–1990 (David B. Wake and Marvalee H. Wake); Curatorial Assistant of Herpetology, 1985, 1989.
Ph.D. dissertation: Development and Evolution in the Salamander Family Plethodontidae.

Cook, Sherburne F., Jr.
Curatorial Assistant of Herpetology, 1950–1952, 1955–1956.

Crippen, Robert G.
Ph.D. student, 1957–1962 (Robert C. Stebbins), degree not completed; Curatorial Assistant of Herpetology, 1958–1963.

Darda, David M.
Ph.D. student, 1980–1988 (David B. Wake); Curatorial Assistant of Herpetology, 1980–1982, 1986.
Ph.D. dissertation: Morphological and Biochemical Evolution within the Plethodontid Salamander Genus *Chiropterotriton*.

Deban, Stephen M.
Ph.D. student, 1991–1997 (David B. Wake); Curatorial Assistant of Herpetology, 1992–1994.
Ph.D. dissertation: Development and Evolution of Feeding Behavior and Functional Morphology in Salamanders of the Family Plethodontidae.

de Queiroz, Alan
Undergraduate Research Assistant of Herpetology, 1979–1981 (Harry W. Greene).

de Queiroz, Kevin
Ph.D. student, 1983–1989 (David B. Wake); Curatorial Assistant of Herpetology, 1983, 1986, 1989.
Ph.D. dissertation: Morphological and Biochemical Evolution in the Sand Lizards.

Dickie, Renée
Ph.D. student, 1993–1999 (Marvalee H. Wake).
Ph.D. dissertation: Structure-Function Relationships in the Evolutionary Morphology of the Plethodontid Tail.

Eaton, Theodore H., Jr.
Ph.D. student, 1930–1933 (Alden H. Miller).
Ph.D. dissertation: Later Stages in the Ontogeny of the Musculature of *Dicamptodon*, with a Comparative Survey of Urodele Myology.

Edwards, James L.
Ph.D. student, 1971–1976 (David B. Wake); Curatorial Assistant of Herpetology, 1970–1971.
Ph.D. dissertation: A Comparative Study of Locomotion in Terrestrial Salamanders.

Elias, Paul
Master's student, 1978–1979 (David B. Wake); Undergraduate Curatorial Assistant of Herpetology, 1974; Curatorial Assistant of Herpetology, 1977–1978.
M.A. thesis: The Salamanders of Northwestern Guatemala.

Endler, John A.
Undergraduate Curatorial Assistant of Herpetology, 1967–1969.

Erickson, Gregory M.
Ph.D. student, 1992–1997 (Marvalee H. Wake).
Ph.D. dissertation: The Evolution of the Biomechanical Attributes of Long Bones.

Etheridge, Richard
Visiting Research Associate, 1975–1976.

Feder, Juliana H.
Master's student, 1976–1977 (James L. Patton and David B. Wake); Curatorial Assistant of Herpetology, 1973–1974.
M.A. thesis: Genetic Variation and Biochemical Systematics in Western *Bufo*.

Feder, Martin E.
Ph.D. student, 1973–1977 (Paul Licht).
Ph.D. dissertation: Bioenergetics of Lungless Salamanders (Caudata: Plethodontidae).

Fellers, Gary M.
Undergraduate Curatorial Assistant of Herpetology, 1970.

Fisher, Edna M.
Assistant Curator of Osteology, 1921–1922, 1923–1930.

Fitch, Henry S.*
Master's student, 1931–1933 (Joseph Grinnell); Ph.D. student, 1933–1937 (J. Grinnell); Technical Assistant, Hastings Natural History Reservation, 1937–1938.
M.A. thesis: Natural History and Systematic Account of the Pacific Coast Alligator Lizards (Genus *Gerrhonotus*).
Ph.D. dissertation: A Biogeographical Study of the *ordinoides* Artenkreis of Garter Snakes (Genus *Thamnophis*).

Fox, R. Wade, Jr.*
Ph.D. student, 1946–1950 (Robert C. Stebbins); Curatorial Assistant of Herpetology, 1943–1949.
Ph.D. dissertation: Biology of the Garter Snakes of the San Francisco Bay Region.

Frelow, Monica M.
Laboratory Technician, 1976–1990.

Fu, Jinzhong
Natural Sciences and Engineering Research Council of Canada Postdoctoral Fellow, 1999–2000.

García–París, Mario
Spanish Ministry of Education and Culture Postdoctoral Fellow, 1993–1994; UC Berkeley Postdoctoral Fellow, 1996–1998.

Glaser, H. S. Robert
Master's student, 1949–1952 (Robert C. Stebbins); Ph.D. student, 1952–1960 (Robert C. Stebbins).
M.A. thesis: Head Form and Differential Growth in the Southwestern Night Lizards, Genus *Xantusia*.
Ph.D. dissertation: The "Third Eye" in Locomotor Activity and the Thermal Ecology of Night Lizards, *Xantusia*.

Good, David A.*
Ph.D. student, 1979–1985 (David B. Wake); Curatorial Assistant of Herpetology, 1984; Curatorial Associate of Herpetology, 1985–1989; Research Associate, 1989–1990.
Ph.D. dissertation: Studies of Interspecific and Intraspecific Variation in the Alligator Lizards (Lacertilia: Anguidae: Gerrhonotinae).

Gorman, George C.
National Institutes of Health Postdoctoral Fellow, 1968; Miller Postdoctoral Fellow, 1969–1971.

Gorman, Joseph B., Jr.
Ph.D. student, 1948–1954 (Robert C. Stebbins).
Ph.D. dissertation: Biosystematic Studies of the Salamanders of the Genus *Hydromantes*.

Graybeal, Anna
Ph.D. student, 1989–1995 (David B. Wake); Curatorial Assistant of Herpetology, 1990.
Ph.D. dissertation: Phylogenetic Relationships and Evolution of Bufonid Frogs Based on Molecular and Morphological Characters.

Green, David M.
Natural Sciences and Engineering Research Council of Canada Postdoctoral Fellow, 1981–1983; Visiting Research Associate, 1983–1984.

Greene, Harry W.*
Assistant Curator of Herpetology, 1978–1983; Associate Curator of Herpetology, 1983–1992; Curator of Herpetology, 1992–1998.

Greer, Allen E.*
Curatorial Associate of Herpetology, 1974–1975.

Gress, Franklin
Curatorial Assistant of Herpetology, 1967–1969.

Gressitt, J. Linsley*
Ph.D. student (Department of Entomology and Parasitology), 1944?–1945.
Ph.D. dissertation: The Tortoise Beetles of China (Chrysomelidae: Cassidinae).

Griffith, Hugh
Natural Sciences and Engineering Research Council of Canada Postdoctoral Fellow, 1991–1993.

Grinnell, Hilda W.*
Curatorial Assistant of Herpetology, 1909.

Grinnell, Joseph*
Director, 1908–1939.

Haddad, Célio F. B.
Visiting Research Associate, 1997.

Hanken, James
Ph.D. student, 1973–1980 (David B. Wake); Curatorial Assistant of Herpetology, 1976.
Ph.D. dissertation: Morphological and Genetic Investigations of Miniaturization in Salamanders (Genus *Thorius*).

Hayes, Tyrone B.*
Ph.D. student, 1989–1993 (Paul Licht); Associate Research Developmental Biologist, 1995–present.
Ph.D. dissertation: The Role of Steroids in Growth and Development in Anuran Amphibians.

Hendrickson, John R.
Ph.D. student, 1944–1952 (Robert C. Stebbins).
Ph.D. dissertation: Studies on the Salamander Genus *Batrachoseps*.

Hetherington, Thomas E.
Ph.D. student, 1974–1979 (Marvalee H. Wake); Curatorial Assistant of Herpetology, 1976.
Ph.D. dissertation: Behavioral and Morphological Analysis of Pineal Organ Function in the Salamander *Ensatina eschscholtzii*.

Highton, Richard
Visiting Research Associate, 1983.

Houck, Lynne D.
Master's student, 1971–1975 (David B. Wake); Ph.D. student, 1975–1977 (David B. Wake); Curatorial Assistant of Herpetology, 1971.
M.A. thesis: Reproductive Biology of a Neotropical Salamander, *Bolitoglossa rostrata*.
Ph.D. dissertation: Reproductive Patterns in Neotropical Salamanders.

Huey, Raymond B.
Undergraduate Research Assistant of Herpetology, 1965–1966 (Robert C. Stebbins); Undergraduate Curatorial Assistant of Herpetology, 1966–1967; Miller Postdoctoral Fellow, 1975–1977.

Jackman, Todd R.
Ph.D. student, 1988–1993 (David B. Wake).
Ph.D. dissertation: Evolutionary and Historical Analyses within and among Members of the Salamander Tribe Plethodontini (Amphibia: Plethodontidae).

Jaksic, Fabián M.
Ph.D. student, 1979–1982 (Harry W. Greene); Curatorial Assistant of Herpetology, 1979–1982.
Ph.D. dissertation: Predation upon Vertebrates in Mediterranean Habitats of Chile, Spain, and California: A Comparative Analysis.

Jockusch, Elizabeth L.
Ph.D. student, 1990–1996 (David B. Wake).
Ph.D. dissertation: Evolutionary Studies in *Batrachoseps* and Other Plethodontid Salamanders: Correlated Character Evolution, Molecular Phylogenetics, and Reaction Norm Evolution.

Johnson, Roy B., Jr.
Master's student, 1937–1939 (Alden H. Miller).
M.A. thesis: Sexual Dimorphism in the Integument of the Fence Lizard.

Jones, Robert E.
Senior Museum Scientist, 1969–present.

Kaplan, Robert H.
Miller Postdoctoral Fellow, 1978–1980; Visiting Research Associate, 1989–1990.

Karlstrom, Ernest L.
Ph.D. student, 1952–1956 (Robert C. Stebbins); Curatorial Assistant of Herpetology, 1953–1954.
Ph.D. dissertation: Ecological and Systematic Relationships within the Toad Genus *Bufo* in the Sierra Nevada of California.

Kirkpatrick, Mark
Miller Postdoctoral Fellow, 1983–1985.

Klosterman, Laurie L.
Ph.D. student, 1977–1982 (Marvalee H. Wake); Curatorial Assistant of Herpetology, 1977–1978.
Ph.D. dissertation: Ultrastructural and Experimental Studies of Oogenesis in the Anguid Lizard *Gerrhonotus coeruleus*.

Koo, Michelle S.–M.
Master's student, 1991–1994 (Marvalee H. Wake); Curatorial Assistant of Herpetology, 1991–1992, 1994.
M.A. thesis: A Description and Evolutionary Interpretation of Dermal Coosification in the Salamander *Aneides lugubris* (Amphibia: Plethodontidae).

Krauss, Max
Master's student, 1938–1940 (Alden H. Miller).
M.A. thesis: The Relation between the Lateral-Line Sense Organs and the Development of Dermal Bones in the Skull of *Triturus torosus* Rathke.

Kuchta, Shawn R.
Ph.D. student, 1996–present (David B. Wake).
Ongoing dissertation research topic: Molecular systematics, speciation, and ecology in the salamander genera *Ensatina* and *Taricha*.

LaPointe, Joseph L.
Ph.D. student, 1961–1966 (Robert C. Stebbins); Curatorial Assistant of Herpetology, 1961, 1963, 1965.
Ph.D. dissertation: Investigation of the Function of the Parietal Eye in Relation to Locomotor Activity Cycles in the Lizard, *Xantusia vigilis*.

Lappin, A. Kristopher
Ph.D. student, 1992–1999 (Marvalee H. Wake).
Ph.D. dissertation: Evolutionary Ecomorphology of the Feeding Biology of Crotaphytid Lizards.

Larson, Allan L.
Ph.D. student, 1977–1982 (David B. Wake).
Ph.D. dissertation: Neontological Inferences of Evolutionary Patterns and Processes in the Salamander Family Plethodontidae.

Larson, Mervin W.
Master's student, 1952–1957 (Robert C. Stebbins).
M.A. thesis: The Critical Thermal Maximum of the Lizard *Sceloporus occidentalis*.

Lawrence, John F.
Curatorial Assistant of Herpetology, 1959–1960.

León, Pedro E.
Visiting Research Associate, 1988.

Lessa, Enrique P.
National Science Foundation Postdoctoral Fellow, 1988–1990.

Linsdale, Jean M.*
Research Associate, 1928–1937; Resident Director, Hastings Natural History Reservation, 1938–1960.

Lombard, R. Eric
Ph.D. student, 1969–1971 (David B. Wake); degree from the University of Chicago. National Science Foundation Postdoctoral Fellow, 1972.

Losos, Jonathan B.
Ph.D. student, 1984–1989 (Harry W. Greene).
Ph.D. dissertation: Ecomorphological Adaptation in the Genus *Anolis*.

Lowe, Charles H., Jr.
Exchange Ph.D. student (from University of California, Los Angeles), 1947–1948; Ph.D. student, 1948–1950 (Robert C. Stebbins). Lowe received his degree from the University of California, Los Angeles.
Ph.D. dissertation: Speciation and Ecology in Salamanders of the Genus *Aneides*.

Luke, Claudia A.
Ph.D. student, 1982–1989 (James L. Patton); Curatorial Assistant of Herpetology, 1985–1988.
Ph.D. dissertation: Color as a Phenotypically Plastic Character in the Side-blotched Lizard, *Uta stansburiana*.

Lynch, James F.*
Ph.D. student, 1967–1974 (David B. Wake); Curatorial Assistant of Herpetology, 1970; Assistant Curator of Herpetology, 1972.
Ph.D. dissertation: Ontogenetic and Geographic Variation in the Morphology and Ecology of the Black Salamander, *Aneides flavipunctatus*.

Macey, J. Robert*
Undergraduate Curatorial Assistant of Herpetology, 1983–1987; Curatorial Assistant of Herpetology, 1989.

Macgregor, Herbert
Visiting Research Associate, 1977–1978.

Mahoney, Meredith J.
Ph.D. student, 1993–1999 (David B. Wake); Curatorial Assistant of Herpetology, 1993–1994; Research Associate, 2000
Ph.D. dissertation: Systematics and Morphological Evolution of *Plethodon* and *Aneides* (Caudata: Plethodontidae: Plethodontini).

Maiorana, Virginia C.
Master's student, 1969–1971 (David B. Wake); Ph.D. student, 1971–1974 (David B. Wake); Curatorial Assistant of Herpetology, 1969–1970.
M.A. thesis: The Foraging Strategy of the California Slender Salamander, *Batrachoseps attenuatus*: An Adaptive Interpretation.
Ph.D. dissertation: Studies in the Behavioral Ecology of the Plethodontid Salamander *Batrachoseps attenuatus*.

Manier, Mollie K.
Undergraduate Curatorial Assistant of Herpetology, 1995, 1996, 1997.

Marks, Sharyn B.
Ph.D. student, 1987–1995 (David B. Wake); Curatorial Assistant of Herpetology, 1987, 1990.
Ph.D. dissertation: Development and Evolution of the Dusky Salamanders (Genus *Desmognathus*).

Marlow, Ronald W.
Ph.D. student, 1972–1979 (Robert C. Stebbins); Curatorial Assistant of Herpetology, 1971–1972, 1978–1979.
Ph.D. dissertation: Energy Relations in the Desert Tortoise, *Gopherus agassizii*.

Maslin, T. Paul*
Master's student, 1936–1939 (E. Raymond Hall); Curatorial Assistant of Herpetology, 1940–1941.
M.A. thesis: A Study of a Collection of Snakes from Lushan, Kiangsi, China.

McGinnis, Samuel M., Jr.
Ph.D. student, 1959–1965 (Robert C. Stebbins).
Ph.D. dissertation: Thermal Ecology of the Western Fence Lizard, *Sceloporus occidentalis.*

Miles, Donald B.
Undergraduate Research Assistant of Herpetology, 1977–1978 (David B. Wake).

Miller, Charles M.
Master's student, 1939–1943 (Alden H. Miller).
M.A. thesis: The Natural History of the Limbless Lizards of the Genus *Anniella.*

Miller, Loye H.
Master's student, 1898–1904 (William E. Ritter).
M.A. thesis: Additional Notes on the Breeding Habits and Development of *Autodax lugubris.*

Moritz, Craig*
Visiting Research Associate, 1990; Curator of Herpetology, 2001–present; Director, 2001–present.

Mueller, Rachel L.
Ph.D. student, 1998–present (David B. Wake).
Ongoing dissertation research topic: The evolution of red blood cell enucleation in the genus *Batrachoseps.*

Mulcahy, Daniel G.
Undergraduate Curatorial Assistant of Herpetology, 1995–1996; Undergraduate Research Assistant of Herpetology, 1997–1998 (Robert C. Stebbins and David B. Wake).

Murray, Keith F.
Ph.D. student, 1948–1952 (Seth B. Benson), degree not completed; Curatorial Assistant of Herpetology, 1949–1952.

Nevo, Eviatar
Visiting Research Associate, 1972–1973.

Nishikawa, Kiisa C.
Miller Postdoctoral Fellow, 1985–1987.

Olson, Wendy M.
Ph.D. student, 1991–1998 (Marvalee H. Wake).
Ph.D. dissertation: Evolutionary and Developmental Morphology of the Dwarf African Clawed Frog, *Hymenochirus boettgeri* (Amphibia: Anura: Pipidae).

Papenfuss, Theodore J.*
Ph.D. student, 1972–1979 (Robert C. Stebbins); Undergraduate Curatorial Assistant of Herpetology, 1959–1960, 1962–1965; Curatorial Assistant of Herpetology, 1966, 1968, 1970, 1979; Research Associate of Herpetology, 1979–present.
Ph.D. dissertation: A Comparative Study of the Ecology and Systematics of the Species within the Amphisbaenian Genus *Bipes*.

Parks, Duncan S.
Ph.D. student, 1994–2000 (David B. Wake).
Ph.D. dissertation: Phylogeography, Historical Distribution, Migration, and Species Boundaries in the Salamander *Ensatina eschscholtzii* as Measured with Mitochondrial DNA Sequences.

Parra–Olea, Gabriela
Ph.D. student, 1994–1999 (David B. Wake).
Ph.D. dissertation: Molecular Evolution and Systematics of Neotropical Salamanders (Caudata: Plethodontidae: Bolitoglossini).

Patton, James L.
Curator of Mammals, 1979–present; Associate Director, 1982–1998; Acting Director, 1988–1989, 1992, 1996, 1998–2000.

Pearson, Anita K.*
Research Associate, 1979–present.

Pearson, Oliver P.*
Assistant Curator of Mammals, 1948–1959; Acting Director, 1966–1967; Director, 1967–1971; Director Emeritus, 1971–present.

Peccinini–Seale, Denise
Visiting Research Associate, 1980–1981.

Poe, Steven
Miller Postdoctoral Fellow, 2000–2002.

Presch, William
Curatorial Assistant of Herpetology, 1972–1973.

Ragghianti, Matilde
Italian Government Postdoctoral Fellow, 1983–1984.

Rainey, William E.
Ph.D. student, 1973–1984 (James L. Patton).
Ph.D. dissertation: Albumin Evolution in Turtles.

Reese, Devin A.
Ph.D. student, 1988–1996 (Harry W. Greene); Curatorial Assistant of Herpetology, 1989–1990.
Ph.D. dissertation: Comparative Demography and Habitat Use of Western Pond Turtles in Northern California: The Effects of Damming and Related Alterations.

Reiserer, Randall S.
Ph.D. student, 1994–2001 (Harry W. Greene and James L. Patton).
Ph.D. dissertation: Evolution of Life Histories in Rattlesnakes.

Riemer, William J.
Ph.D. student, 1949–1956 (Robert C. Stebbins); Undergraduate Curatorial Assistant of Herpetology, 1947–1949; Curatorial Assistant of Herpetology, 1949–1954.
Ph.D. dissertation: Variation and Systematic Relationships within the Salamander Genus *Taricha*.

Riney, Thane A.
Master's student, 1946–1951 (A. Starker Leopold).
M.A. thesis: Home Range and Seasonal Movement in a Sierra Deer Herd.

Rissler, Leslie J.
National Science Foundation Postdoctoral Fellow, 2001–present.

Roberts, Wendy E.
Ph.D. student, 1988–1994 (Harry W. Greene).
Ph.D. dissertation: Evolution and Ecology of Arboreal Egg-laying Frogs.

Rodgers, Thomas L.*
Master's student, 1935–1937 (Alden H. Miller); Ph.D. student, 1938?–1953 (A. H. Miller); Curatorial Assistant of Herpetology, 1936–1937; Acting Curator of Herpetology, 1937–1945.
M.A. thesis: A Study of Variation in the Desert Whip–tailed Lizard, *Cnemidophorus tesselatus* Say.
Ph.D. dissertation: Responses of Two Closely Related Species of Lizards (Genus *Sceloporus*) to Different Environmental Conditions.

Rodríguez–Robles, Javier A.*
Ph.D. student, 1992–1998 (Harry W. Greene); Curatorial Assistant of Herpetology, 1993–1995, 1996, 1997, 1998; Acting Assistant Curator of Herpetology, 1999–2000; National Science Foundation Postdoctoral Fellow, 2000–2002.
Ph.D. dissertation: Molecular Systematics and Feeding Ecology of

Lampropeltinine Snakes.

Rosenblum, Erica B.
Ph.D. student, 2000–present (Craig Moritz and David B. Wake).
Ongoing dissertation research topic: Local adaptation, environmental gradients, and the genetics of color variation in lizards from White Sands National Monument, New Mexico.

Rosenthal, Gerson M.
Ph.D. student, 1948–1954 (Robert C. Stebbins).
Ph.D. dissertation: The Role of Moisture and Temperature in the Local Distribution of the Plethodontid Salamander *Aneides lugubris*.

Roth, Gerhard
Visiting Research Associate, 1974, 1982.

Ruben, John A.
Ph.D. student, 1973–1975 (Paul Licht).
Ph.D. dissertation: Physiological and Morphological Correlates of Activity Modes in Snakes.

Russell, Ward C.
Museum Zoologist, 1929–1969.

Ruth, Stephen B.
Ph.D. student, 1970–1977 (Robert C. Stebbins); Curatorial Assistant of Herpetology, 1969–1975, 1977.
Ph.D. dissertation: A Comparison of the Demography and Female Reproduction in Sympatric Western Fence Lizards (*Sceloporus occidentalis*) and Sagebrush Lizards (*Sceloporus graciosus*) on Mount Diablo, California.

Ryan, Michael J.
Miller Postdoctoral Fellow, 1982–1984.

Sage, Richard D.*
Undergraduate Curatorial Assistant of Herpetology, 1963–1966; Curatorial Associate of Herpetology, 1976–1978, 1986; Curatorial Assistant of Herpetology, 1983–1984.

Schneider, Christopher J.
Ph.D. student, 1987–1993 (Harry W. Greene and David B. Wake); Curatorial Assistant of Herpetology, 1988, 1993.
Ph.D. dissertation: Diversification in Lizards of the Genus *Anolis* from Guadeloupe and the Northern Lesser Antilles.

Schuler, Alexis L.
Undergraduate Curatorial Assistant of Herpetology, 1987–1988.

Schwenk, Kurt
Ph.D. student, 1978–1984 (Marvalee H. Wake); Curatorial Assistant of Herpetology, 1983.
Ph.D. dissertation: Evolutionary Morphology of the Lepidosaur Tongue.

Seeliger, Lillian M.
Ph.D. student, 1944 (no major professor), degree not completed.

Seib, Robert L.*
Ph.D. student, 1979–1985 (Harry W. Greene); Curatorial Assistant of Herpetology, 1977–1985.
Ph.D. dissertation: Feeding Ecology and Organization of Neotropical Snake Faunas.

Sessions, Stanley K.
Ph.D. student, 1978–1985 (David B. Wake).
Ph.D. dissertation: Cytogenetics and Evolution in Salamanders.

Shaffer, H. Bradley
Undergraduate Curatorial Assistant of Herpetology, 1974–1976; Curatorial Assistant of Herpetology, 1976–1977; Visiting Graduate Student, 1980–1981.

Sheen, Judy P.
Ph.D. student, 1996–2001 (Marvalee H. Wake).
Ph.D. dissertation: Reproductive cost in the northern and southern alligator lizards (*Elgaria coerulea* and *E. multicarinata*).

Sherman, Paul W.
Miller Postdoctoral Fellow, 1976–1978.

Sinervo, Barry
Miller Postdoctoral Fellow, 1988–1990.

Spoeker, Peter D.
Ph.D. student, 1964–1966 (Robert C. Stebbins), degree not completed.

Stamps, Judy A.
Master's student, 1969–1971 (George W. Barlow); Ph.D. student, 1971–1974 (George W. Barlow).
M.A. thesis: Variation and Stereotypy in *Anolis* Display.
Ph.D. dissertation: Spacing Patterns in Lizards.

Staub, Nancy L.
Ph.D. student, 1981–1989 (David B. Wake); Curatorial Assistant of Herpetology, 1984, 1986–1988; 2000–2001; Visiting Research Associate, 2000–2001.
Ph.D. dissertation: The Evolution of Sexual Dimorphism in the Salamander Genus *Aneides* (Amphibia: Plethodontidae).

Stebbins, Robert C.*
Curator of Herpetology, 1945–1978; Curator Emeritus of Herpetology, 1978–present.

Stewart, James R.
Ph.D. student, 1971–1976 (Robert C. Stebbins and David B. Wake); Curatorial Assistant of Herpetology, 1971–1976.
Ph.D. dissertation: Interpopulational Variation in the Demography of a Live Bearing Lizard *Gerrhonotus coeruleus*.

Storer, Tracy I.*
Master's student, 1912–1913 (Joseph Grinnell); Ph.D. student, 1913?–1925 (J. Grinnell); Field Naturalist, 1917–1923.
M.S. thesis: The Toads of California of the Genus *Bufo*.
Ph.D. dissertation: The Amphibia of California: A Synopsis, with Life Histories of the Principal Species Considered in Relation to the Semiarid Climate of the American Southwest.

Summers, Adam P.
Miller Postdoctoral Fellow, 1999–2001.

Sweet, Samuel S.*
Master's student, 1970–1973 (David B. Wake); Ph.D. student, 1973–1978 (David B. Wake); Curatorial Assistant of Herpetology, 1970–1972, 1974–1975, 1977.
M.A. thesis: Allometry, Life History, and the Evolution of the Desmognathine Salamanders.
Ph.D. dissertation: The Evolutionary Development of the Texas *Eurycea* (Amphibia: Plethodontidae).

Tan, An-Ming
Ph.D. student, 1988–1993 (David B. Wake); Curatorial Assistant of Herpetology, 1988–1989.
Ph.D. dissertation: Systematics, Phylogeny and Biogeography of the Northwest American Newts of the Genus *Taricha* (Caudata: Salamandridae).

Taylor, Emily N.
Undergraduate Curatorial Assistant of Herpetology, 1998–1999.

Tevis, Lloyd P., Jr.
Research Assistant, Hastings Natural History Reservation, 1941–1948.

Theodoratus, Demetri H.
Master's student, 1995–1998 (Harry W. Greene).
M.A. thesis: Studies on the Foraging Ecology of Crotaline Snakes.

Tillotson, Daniel F.
Master's student, 1938–1941 (E. Raymond Hall).
M.A. thesis: A Biometric Analysis of the Variation in the Lizard, *Uta stansburiana hesperis*.

Tollestrup, Kristine
Ph.D. student, 1973–1979 (Robert C. Stebbins); Curatorial Assistant of Herpetology, 1977–1978.
Ph.D. dissertation: The Ecology, Social Structure and Foraging Behavior of Two Closely Related Species of Leopard Lizards, *Gambelia silus* and *Gambelia wislizenii*.

Troyer, Katherine
Research Associate, 1982–1984.

Turner, Frederick B.
Ph.D. student, 1949–1957 (Robert C. Stebbins).
Ph.D. dissertation: The Ecology and Morphology of *Rana pretiosa pretiosa* in Yellowstone Park, Wyoming.

Volz, Eugene R.
Master's student, 1954–1957 (Robert C. Stebbins); Curatorial Assistant of Herpetology, 1954–1957.
M.A. thesis: The Sexual Cycle and Reproductive Anatomy of *Gerrhonotus coeruleus*, the Northern Alligator Lizard.

Vredenburg, Vance T.
Ph.D. student, 1995–present (Mary E. Power and David B. Wake).
Ongoing dissertation research topic: Ecology and conservation of the mountain yellow-legged frog, *Rana muscosa*.

Wake, David B.*
Associate Curator of Herpetology, 1969–1971; Curator of Herpetology, 1971–present; Director, 1971–1998.

Wake, Marvalee H.*
Research Morphologist, 1975–present.

Wassersug, Richard J.
Ph.D. student, 1969–1973 (Robert F. Inger and David B. Wake). Wassersug received his degree from the University of Chicago. Visiting Research Associate, 1986–1987.

White, Marshall
Associate Research Conservationist, 1968–1989.

Wilhoft, Daniel C.
Master's student, 1956–1958 (Robert C. Stebbins); Ph.D. student, 1958–1963 (Robert C. Stebbins); Curatorial Assistant of Herpetology, 1957–1958.
M.A. thesis: The Effect of Temperature on Thyroid Histology and Survival in the Lizard *Sceloporus occidentalis*.
Ph.D. dissertation: Seasonal Changes in the Gonads and Thyroid in a Tropical Lizard, *Leiolopisma rhomboidalis*.

Wood, Wallace F.
Undergraduate student, 1932–1936.

Wurst, Gloria Z.
Curatorial Assistant of Herpetology, 1977.

Yanev, Kay P.*
Ph.D. student, 1971–1978 (David B. Wake); Research Associate, 1979–1985.
Ph.D. dissertation: Evolutionary Studies of the Plethodontid Salamander Genus *Batrachoseps*.

Yang, Suh Y.*
Curatorial Associate, 1973–1976.

Zamudio, Kelly R.
Undergraduate Curatorial Assistant of Herpetology, 1991; National Science Foundation Postdoctoral Fellow, 1997–1998.

Zweifel, Richard G.
Ph.D. student, 1950–1954 (Robert C. Stebbins); Curatorial Assistant of Herpetology, 1951–1953.
Ph.D. dissertation: Ecology, Distribution and Systematics in the *boylei* Group of the Genus *Rana*.

Literature Cited

Anonymous
1879   Proceedings of scientific societies. Amer. Nat. 13:799-801.

1993   Opinion 1735. *Anniella pulchra* Gray, 1852 (Reptilia, Squamata): neotype designated. Bull. Zool. Nomencl. 50:186-187.

1997   Opinion 1866. *Hydromantes* Gistel, 1848 (Amphibia, Caudata): *Spelerpes platycephalus* Camp, 1916 designated as the type species. Bull. Zool. Nomencl. 54:72-74.

Adler, K.
1996   The salamanders of Guerrero, Mexico, with descriptions of five new species of *Pseudoeurycea* (Caudata: Plethodontidae). Occ. Pap. Nat. Hist. Mus., Univ. Kansas 177:1-28.

Adler, K. K., and D. M. Dennis
1962   *Plethodon longicrus,* a new salamander (Amphibia: Plethodontidae) from North Carolina. Ohio Herpetol. Soc. Spec. Publ. 4:1-14.

Alberch, P., S. J. Gould, G. F. Oster, and D. B. Wake
1979   Size and shape in ontogeny and phylogeny. Paleobiology 5:296-317.

Alexander, A. M.
1907   Letter to Benjamin Ide Wheeler (manuscript CU-5), Oakland, California, October 28, 1907. Located at Records of the Office of the

President, University Archives, Bancroft Library, Univ. California, Berkeley.

Anderson, J. D.
    1962   A new subspecies of the ridged-nosed rattlesnake, *Crotalus willardi*, from Chihuahua, Mexico. Copeia 1962:160-163.

Anderson, J. D., and D. C. Wilhoft
    1959   A new subspecies of *Eumeces multivirgatus* from Mexico. Copeia 1959:57-60.

Atsatt, S. R.
    1913   The reptiles of the San Jacinto area of southern California. Univ. California Publ. Zool. 12:31-50.

Barbour, T., and A. Loveridge
    1928   A comparative study of the herpetological faunae of the Uluguru and Usambara Mountains, Tanganyika Territory with descriptions of new species. Mem. Mus. Comp. Zool., Harvard College 50:87-265.

Beaman, K. R., and L. L. Grismer
    1994   *Crotalus enyo*. Cat. Amer. Amphib. Rept. 589.1-589.6.

Bell, E. L.
    1954   A preliminary report on the subspecies of the western fence lizard, *Sceloporus occidentalis*, and its relationships to the eastern fence lizard, *Sceloporus undulatus*. Herpetologica 10:31-36.

Blanchard, F. N.
    1923   Comments on ring-neck snakes (genus *Diadophis*), with diagnoses of new forms. Occ. Pap. Mus. Zool., Univ. Michigan 142:1-9.

Bolaños, F., D. C. Robinson, and D. B. Wake
    1987   A new species of salamander (genus *Bolitoglossa*) from Costa Rica. Rev. Biol. Trop. 35:87-92.

Boundy, J.
    1999   Systematics of the garter snake *Thamnophis atratus* at the southern end of its range. Proc. California Acad. Sci. 51:311-336.

Brame, A. H., Jr., and K. F. Murray
    1968   Three new slender salamanders (*Batrachoseps*) with a discussion of relationships and speciation within the genus. Bull. Nat. Hist. Mus. Los Angeles County 4:1-35.

Brame, A. H., Jr., and D. B. Wake
  1971 New species of salamanders (genus *Bolitoglossa*) from Colombia, Ecuador, and Panamá. Contrib. Sci., Nat. Hist. Mus. Los Angeles County 219:1-34.

Brown, W. C., and J. T. Marshall Jr.
  1953 New scincoid lizards from the Marshall Islands, with notes on their distribution. Copeia 1953:201-207.

Busack, S. D.
  1988 Biochemical and morphological differentiation in Spanish and Moroccan populations of *Blanus* and the description of a new species from northern Morocco (Reptilia, Amphisbaenia, Amphisbaenidae). Copeia 1988:101-109.

Cadle, J. E.
  1991 Systematics of lizards of the genus *Stenocercus* (Iguania: Tropiduridae) from northern Perú: New species and comments on relationships and distribution patterns. Proc. Acad. Nat. Sci. Philadelphia 143:1-96.

  1998 New species of lizards, genus *Stenocercus* (Iguania: Tropiduridae), from western Ecuador and Peru. Bull. Mus. Comp. Zool. 155:257-297.

Camp, C. L.
  1915 *Batrachoseps major* and *Bufo cognatus californicus*, new Amphibia from southern California. Univ. California Publ. Zool. 12:327-334.

  1916a *Spelerpes platycephalus*, a new alpine salamander from the Yosemite National Park, California. Univ. California Publ. Zool. 17:11-14.

  1916b Description of *Bufo canorus*, a new toad from the Yosemite National Park. Univ. California Publ. Zool. 17:59-62.

  1916c The subspecies of *Sceloporus occidentalis* with description of a new form from the Sierra Nevada and systematic notes on other California lizards. Univ. California Publ. Zool. 17:63-74.

  1917 Notes on the systematic status of the toads and frogs of California. Univ. California Publ. Zool. 17:115-125.

Campbell, J. A.
  1985 A new species of highland pitviper of the genus *Bothrops* from southern Mexico. J. Herpetol. 19:48-54.

Campbell, J. A., and B. T. Clarke
    1998   A review of frogs of the genus *Otophryne* (Microhylidae) with the description of a new species. Herpetologica 54:301-317.

Campbell, J. A., and D. R. Frost
    1993   Anguid lizards of the genus *Abronia*: Revisionary notes, descriptions of four new species, a phylogenetic analysis, and key. Bull. Amer. Mus. Nat. Hist. 216:1-121.

Campbell, J. A., and J. M. Savage
    2000   Taxonomic reconsideration of Middle American frogs of the *Eleutherodactylus rugulosus* group (Anura: Leptodactylidae): A reconnaissance of subtle nuances among frogs. Herpetol. Monogr. 14:186-292.

Campbell, J. A., and E. N. Smith
    1998   A new genus and species of colubrid snake from the Sierra de las Minas of Guatemala. Herpetologica 54:207-220.

Chippindale, P. T., A. H. Price, J. J. Wiens, and D. M. Hillis
    2000   Phylogenetic relationships and systematic revision of central Texas hemidactyliine plethodontid salamanders. Herpetol. Monogr. 14:1-80.

Cope, E. D.
    1884 (1883)   Notes on the geographical distribution of Batrachia and Reptilia in western North America. Proc. Acad. Nat. Sci. Philadelphia 35:10-35.

Crippen, R. G.
    1962   Holotype specimens of amphibians and reptiles in the Museum of Vertebrate Zoology, University of California, Berkeley. Herpetologica 18:187-194.

Dixon, J. R.
    1964   The systematics and distribution of lizards of the genus *Phyllodactylus* in North and Central America. New Mexico State Univ. Res. Center Sci. Bull. 64-1:1-139.

    1983   The *Liophis cobella* group of the Neotropical colubrid snake genus *Liophis*. J. Herpetol. 17:149-165.

Dixon, J. R., and R. B. Huey
    1970   Systematics of the lizards of the gekkonid genus *Phyllodactylus* of mainland South America. Contrib. Sci., Nat. Hist. Mus. Los Angeles County 192:1-78.

Eakin, R. M.
  1956 History of Zoology at the University of California, Berkeley. Bios 27:67-92.

  1988 History of Zoology at Berkeley. Dept. Zool., Univ. California, Berkeley. 91 pp.

Eakin, R. M., and R. C. Stebbins
  1959 Parietal eye nerve in the fence lizard. Science 130:1573-1574.

Elias, P.
  1984 Salamanders of the northwestern highlands of Guatemala. Contrib. Sci., Nat. Hist. Mus. Los Angeles County 348:1-20.

Elias, P., and D. B. Wake
  1983 *Nyctanolis pernix,* a new genus and species of plethodontid salamander from northwestern Guatemala and Chiapas, Mexico. Pp. 1-12 in Advances in Herpetology and Evolutionary Biology: Essays in Honor of Ernest E. Williams (G. J. Rhodin and K. Miyata, eds.). Mus. Comp. Zool., Cambridge, Massachusetts.

Ernst, C. H.
  1984 (1983) Geographic variation in the Neotropical turtle, *Platemys platycephala.* J. Herpetol. 17:345-355.

Esterly, C. O.
  1904 The structure and regeneration of the poison glands of *Plethodon.* Univ. California Publ. Zool. 1:227-268.

Estes, R.
  1988 Charles Camp: An appreciation. Pp. 9-14 in Phylogenetic Relationships of the Lizard Families: Essay Commemorating Charles L. Camp (R. Estes and G. Pregill, eds.). Stanford Univ. Press, Stanford, California.

Fitch, H. S.
  1934a New alligator lizards from the Pacific Coast. Copeia 1934:6-7.

  1934b A shift of specific names in the genus *Gerrhonotus.* Copeia 1934:172-173.

  1936 Amphibians and reptiles of the Rogue River Basin, Oregon. Amer. Midl. Nat. 17:634-652.

  1940 A biogeographical study of the *ordinoides* artenkreis of garter snakes (genus *Thamnophis*). Univ. California Publ. Zool. 44: 1-149.

Fitch, H. S.
 1999 A Kansas Snake Community: Composition and Changes over 50 Years. Krieger, Malabar, Florida. 165 pp.

Fitch, H. S. (as told to Alice Fitch Echelle)
 2000 Historical perspective: Henry S. Fitch. Copeia 2000:891-900.

Ford, L. S., and J. M. Savage
 1984 A new frog of the genus *Eleutherodactylus* (Leptodactylidae) from Guatemala. Occ. Pap. Nat. Hist. Mus., Univ. Kansas 110:1-9.

Fox, W.
 1951a Relationships among the garter snakes of the *Thamnophis elegans* rassenkreis. Univ. California Publ. Zool. 50:485-529.

 1951b The status of the gartersnake, *Thamnophis sirtalis tetrataenia*. Copeia 1951:257-267.

 1954 Genetic and environmental variation in the timing of the reproductive cycles of male garter snakes. J. Morphol. 95:415-450.

García-París, M., and D. B. Wake
 2000 Molecular phylogenetic analysis of relationships of the tropical salamander genera *Oedipina* and *Nototriton*, with descriptions of a new genus and three new species. Copeia 2000:42-70.

Good, D. A., and K. Schwenk
 1985 A new species of *Abronia* (Lacertilia: Anguidae) from Oaxaca, Mexico. Copeia 1985:135-141.

Good, D. A., and D. B. Wake
 1992 Geographic variation and speciation in the torrent salamanders of the genus *Rhyacotriton* (Caudata: Rhyacotritonidae). Univ. California Publ. Zool. 126:1-91.

 1993 Systematic studies of the Costa Rican moss salamanders, genus *Nototriton*, with descriptions of three new species. Herpetol. Monogr. 7:131-159.

Gorman, J.
 1954 A new species of salamander from central California. Herpetologica 10:153-158.

 1960 Treetoad studies, 1. *Hyla californiae*, new species. Herpetologica 16:214-222.

Gorman, J., and C. L. Camp
　　1953　A new cave species of salamander of the genus *Hydromantes* from California, with notes on habits and habitat. Copeia 1953: 39-43.

Greene, H. W.
　　1986　Diet and arboreality in the emerald monitor, *Varanus prasinus*, with comments on the study of adaptation. Fieldiana, Zool. (New Ser.) 31:1-12.

　　1992　The ecological and behavioral context for pitviper evolution. Pp. 107-117 in Biology of the Pitvipers (J. A. Campbell and E. D. Brodie Jr., eds.). Selva, Tyler, Texas.

　　1997　Snakes: The Evolution of Mystery in Nature. Univ. California Press, Berkeley. 351 pp.

Greer, A. E., and M. Simon
　　1982　*Fojia bumui*, an unusual new genus and species of scincid lizard from New Guinea. J. Herpetol. 16:131-139.

Gressitt, J. L.
　　1936　New reptiles from Formosa and Hainan. Proc. Biol. Soc. Washington 49:117-121.

　　1937　A new snake from southeastern China. Proc. Biol. Soc. Washington 50:125-128.

　　1938a　A new burrowing frog and a new lizard from Hainan Island. Proc. Biol. Soc. Washington 51:127-130.

　　1938b　Some amphibians from Formosa and the Ryu Kyu Islands, with description of a new species. Proc. Biol. Soc. Washington 51: 159-164.

Grinnell, H. W.
　　1940　Joseph Grinnell: 1877-1939. Condor 42:3-34.

　　1958　Annie Montague Alexander. Grinnell Naturalists Soc., Berkeley, California. 27 pp.

Grinnell, J.
　　1910　The methods and uses of a research museum. Popular Sci. Mon. 77(August):163-169.

　　1914　An account of the mammals and birds of the lower Colorado Valley with especial reference to the distributional problems presented. Univ. California Publ. Zool. 12:51-294.

Grinnell, J.
  1933 Review of the recent mammal fauna of California. Univ. California Publ. Zool. 40:71-234.

Grinnell, J., and C. L. Camp
  1917 A distributional list of the amphibians and reptiles of California. Univ. California Publ. Zool. 17:127-208.

Grinnell, J., J. Dixon, and J. M. Linsdale
  1930 Vertebrate natural history of a section of northern California through the Lassen Peak region. Univ. California Publ. Zool. 35:1-594.

Grinnell, J., and H. W. Grinnell
  1907 Reptiles of Los Angeles County, California. Throop Inst. Bull. (Science Ser. 1) 35:1-64.

Grinnell, J., and A. H. Miller
  1944 The Distribution of the Birds of California. The [Cooper Ornithological] Club, Berkeley, California. 608 pp.

Grinnell, J., and T. I. Storer
  1924 Animal Life in the Yosemite: An Account of the Mammals, Birds, Reptiles, and Amphibians in a Cross-Section of the Sierra Nevada. Univ. California Press, Berkeley. 752 pp.

Grismer, L. L.
  1994 The origin and evolution of the peninsular herpetofauna of Baja California, México. Herpetol. Nat. Hist. 2:51-106.

Hall, W. P., and H. M. Smith
  1979 Lizards of the *Sceloporus orcutti* complex of the Cape Region of Baja California. Breviora 452:1-26.

Hanken, J., and D. B. Wake
  1994 Five new species of minute salamanders, genus *Thorius* (Caudata: Plethodontidae), from northern Oaxaca, Mexico. Copeia 1994:573-590.

  1998 Biology of tiny animals: Systematics of the minute salamanders (*Thorius*: Plethodontidae) from Veracruz and Puebla, México, with descriptions of five new species. Copeia 1998:312-345.

  2001 A seventh species of minute salamander (*Thorius*: Plethodontidae) from the Sierra de Juárez, Oaxaca, México. Herpetologica 57:515-523.

Hanken, J., D. B. Wake, and H. L. Freeman
  1999 Three new species of minute salamanders (*Thorius*: Plethodontidae) from Guerrero, México, including the report of a novel dental polymorphism in urodeles. Copeia 1999:917-931.

Highton, R., and A. H. Brame Jr.
  1965 *Plethodon stormi* species nov. Pilot Register of Zoology, Card No. 20.

Highton, R., and A. B. Grobman
  1956 Two new salamanders of the genus *Plethodon* from the southeastern United States. Herpetologica 12:185-188.

Highton, R., and R. D. Worthington
  1967 A new salamander of the genus *Plethodon* from Virginia. Copeia 1967:617-626.

Hillis, D. M., and J. S. Frost
  1985 Three new species of leopard frogs (*Rana pipiens* complex) from the Mexican Plateau. Occ. Pap. Nat. Hist. Mus., Univ. Kansas 117:1-14.

Holmes, S. J.
  1906 The Biology of the Frog. Macmillan, New York. 370 pp.

Hubbard, M. E.
  1903 Correlated protective devices in some California salamanders. Univ. California Publ. Zool. 1:157-170.

Huey, R. B., and J. R. Dixon
  1970 A new *Pseudogonatodes* from Peru with remarks on other species of the genus. Copeia 1970:538-542.

Hunt, L. E.
  1983 A nomenclature rearrangement of the genus *Anniella* (Sauria: Anniellidae). Copeia 1983:79-89.

Inger, R. F., N. Orlov, and I. Darevsky
  1999 Frogs of Vietnam: A report on new collections. Fieldiana, Zool. (New Ser.) 92:1-46.

Ingram, W., III, and W. W. Tanner
  1971 A taxonomic study of *Crotaphytus collaris* between the Rio Grande and Colorado Rivers. Brigham Young Univ. Sci. Bull., Biol. Ser. 13:1-29.

International Commission on Zoological Nomenclature
  1999   International Code of Zoological Nomenclature, 4th. ed. International Trust for Zoological Nomenclature, London. 306 pp.

Ipsen, D. C., R. C. Stebbins, and G. Gillfillan
  1966a   Animal Coloration, an Introduction to Natural Selection. Part II. Advertising Coloration. Elementary School Science Project, Univ. California, Berkeley. 35 pp.

  1966b   Animal Colors That Advertise. Elementary School Science Project, Univ. California, Berkeley. 42 pp.

Jackman, T. R.
  1999 (1998)   Molecular and historical evidence for the introduction of clouded salamanders (genus *Aneides*) to Vancouver Island, British Columbia, Canada, from California. Can. J. Zool. 76:1570-1580.

Jackman, T. R., G. Applebaum, and D. B. Wake
  1997   Phylogenetic relationships of bolitoglossine salamanders: A demonstration of the effects of combining morphological and molecular data sets. Mol. Biol. Evol. 14:883-891.

Jockusch, E. L., D. B. Wake, and K. P. Yanev
  1998   New species of slender salamanders, *Batrachoseps* (Amphibia: Plethodontidae), from the Sierra Nevada of California. Contrib. Sci., Nat. Hist. Mus. Los Angeles County 472:1-17.

Jockusch, E. L., K. P. Yanev, and D. B. Wake
  2001   Molecular phylogenetic analysis of slender salamanders, genus *Batrachoseps* (Amphibia: Plethodontidae), from central coastal California with descriptions of four new species. Herpetol. Monogr. 15:54-99.

Johnson, M. L.
  1947   The status of the *elegans* subspecies of *Thamnophis*, with description of a new subspecies from Washington State. Herpetologica 3:159-165.

Johnson, N. K.
  1995   Ornithology at the Museum of Vertebrate Zoology, University of California, Berkeley. Pp. 183-221 in Contributions to the History of North American Ornithology (W. E. Davis Jr. and J. A. Jackson, eds.). Memoirs of the Nuttall Ornithological Club, No. 12, Cambridge, Massachusetts.

Karges, J. P., and J. W. Wright
    1987   A new species of *Barisia* (Sauria, Anguidae) from Oaxaca, Mexico. Contrib. Sci., Nat. Hist. Mus. Los Angeles County 381:1-11.

Klauber, L. M.
    1930   New and renamed subspecies of *Crotalus confluentus* Say, with remarks on related species. Trans. San Diego Soc. Nat. Hist. 6:95-144.

    1941   The long-nosed snakes of the genus *Rhinocheilus*. Trans. San Diego Soc. Nat. Hist. 9:289-330.

    1945   The geckos of the genus *Coleonyx* with descriptions of new subspecies. Trans. San Diego Soc. Nat. Hist. 10:133-213.

    1949a Some new and revived subspecies of rattlesnakes. Trans. San Diego Soc. Nat. Hist. 11:61-110.

    1949b The subspecies of the ridge-nosed rattlesnake, *Crotalus willardi*. Trans. San Diego Soc. Nat. Hist. 11:121-137.

Kluge, A. G.
    1993   Gekkonoid Lizard Taxonomy. International Gecko Society, San Diego, California. 245 pp.

Lanza, B., and S. Vanni
    1981   On the biogeography of plethodontid salamanders (Amphibia Caudata) with a description of a new genus. Monitore Zool. Ital. (n.s.) 15:117-121.

Lathrop, A., R. W. Murphy, N. Orlov, and C. T. Ho
    1998   Two new species of *Leptolalax* (Anura: Megophryidae) from northern Vietnam. Amphibia-Reptilia 19:253-267.

Leonard, W. P., and R. C. Stebbins
    1999   Observations of antipredator tactics of the sharp-tailed snake (*Contia tenuis*). Northwest. Nat. 80:74-77.

Leviton, A. E., S. C. Anderson, K. Adler, and S. A. Minton
    1992   Handbook to Middle East Amphibians and Reptiles. Contrib. Herpetol., No. 12. Soc. Stud. Amphib. Rept. 252 pp.

Linsdale, J. M.
    1932   Amphibians and reptiles from Lower California. Univ. California Publ. Zool. 38:345-386.

Linsdale, J. M.
 1940 Amphibians and reptiles in Nevada. Proc. Amer. Acad. Arts Sci. 73:197-257.

 1943 Work with vertebrate animals on the Hastings Natural History Reservation. Amer. Midl. Nat. 30:254-267.

Lombard, R. E., and D. B. Wake
 1977 Tongue evolution in the lungless salamanders, Family Plethodontidae. II. Function and evolutionary diversity. J. Morphol. 153:39-79.

Losos, J. B., and H. W. Greene
 1988 Ecological and evolutionary implications of diet in monitor lizards. Biol. J. Linn. Soc. 35:379-407.

Lowe, C. H., Jr.
 1955 The salamanders of Arizona. Trans. Kansas Acad. Sci. 58:237-251.

Lowe, C. H., Jr., and K. S. Norris
 1954 Analysis of the herpetofauna of Baja California, Mexico. Trans. San Diego Soc. Nat. Hist. 12:47-64.

Lowe, C. H., Jr., and R. G. Zweifel
 1952 A new species of whiptailed lizard (genus *Cnemidophorus*) from New Mexico. Bull. Chicago Acad. Sci. 9:229-247.

Lynch, J. F., and D. B. Wake
 1975 Systematics of the *Chiropterotriton bromeliacia* group (Amphibia: Caudata), with description of two new species from Guatemala. Contrib. Sci., Nat. Hist. Mus. Los Angeles County 265:1-45.

 1978 A new species of *Chiropterotriton* (Amphibia: Caudata) from Baja Verapaz, Guatemala, with comments on relationships among Central American members of the genus. Contrib. Sci., Nat. Hist. Mus. Los Angeles County 294:1-22.

 1989 Two new species of *Pseudoeurycea* (Amphibia: Caudata) from Oaxaca, Mexico. Contrib. Sci., Nat. Hist. Mus. Los Angeles County 411:11-22.

Lynch, J. F., D. B. Wake, and S. Y. Yang
 1983 Genic and morphological differentiation in Mexican *Pseudoeurycea* (Caudata: Plethodontidae), with a description of a new species. Copeia 1983:884-894.

Marlow, R. W., J. M. Brode, and D. B. Wake
   1979   A new salamander, genus *Batrachoseps*, from the Inyo Mountains of California, with a discussion of relationships in the genus. Contrib. Sci., Nat. Hist. Mus. Los Angeles County 308:1-17.

McCranie, J. R., and G. A. Cruz
   1996   A new species of salamander of the *Bolitoglossa dunni* group (Caudata: Plethodontidae) from the Sierra de Agalta, Honduras. Carib. J. Sci. 32:195-200.

McCranie, J. R., and G. Köhler
   1999   A new species of salamander of the *Bolitoglossa dunni* group from Cerro El Pital, Honduras and El Salvador (Amphibia, Caudata, Plethodontidae). Senckenberg. Biol. 78:225-229.

McCranie, J. R., D. B. Wake, and L. D. Wilson
   1996   The taxonomic status of *Bolitoglossa schmidti*, with comments on the biology of the Mesoamerican salamander *Bolitoglossa dofleini* (Caudata: Plethodontidae). Carib. J. Sci. 32:395-398.

McCranie, J. R., and L. D. Wilson
   1993   A review of the *Bolitoglossa dunni* group (Amphibia: Caudata) from Honduras with the description of three new species. Herpetologica 49:1-15.

   1995   A new species of salamander of the genus *Bolitoglossa* (Caudata: Plethodontidae) from Parque Nacional El Cusuco, Honduras. J. Herpetol. 29:447-452.

   1997   A review of the *Eleutherodactylus milesi*-like frogs (Anura, Leptodactylidae) from Honduras with the description of four new species. Alytes 14:147-174.

McCranie, J. R., L. D. Wilson, and J. Polisar
   1998   Another new montane salamander (Amphibia: Caudata: Plethodontidae) from Parque Nacional Santa Barbara, Honduras. Herpetologica 54:455-461.

McDiarmid, R. W., J. A. Campbell, and T. A. Touré
   1999   Snake Species of the World. A Taxonomic and Geographic Reference, Vol. 1. Herpetologists' League, Washington, D.C. 511 pp.

McGuire, J. A.
   1996   Phylogenetic systematics of crotaphytid lizards (Reptilia: Iguania: Crotaphytidae). Bull. Carnegie Mus. Nat. Hist. 32:1-143.

Mendelson, J. R., III
    1997a A new species of toad (Anura: Bufonidae) from the Pacific highlands of Guatemala and southern Mexico, with comments on the status of *Bufo valliceps macrocristatus*. Herpetologica 53:14-30.

    1997b A new species of toad (Anura: Bufonidae) from Oaxaca, Mexico with comments on the status of *Bufo cavifrons* and *Bufo cristatus*. Herpetologica 53:268-286.

Mittleman, M. B.
    1941 A new lizard of the genus *Uta* from Arizona. Proc. Biol. Soc. Washington 54:165-167.

Mittleman, M. B., and G. S. Myers
    1949 Geographic variation in the ribbed frog, *Ascaphus truei*. Proc. Biol. Soc. Washington 62:57-66.

Murphy, R. W.
    1983 Paleobiogeography and genetic differentiation of the Baja California herpetofauna. Occ. Pap. California Acad. Sci. 137:1-48.

Murphy, R. W., and H. M. Smith
    1991 *Anniella pulchra* Gray, 1852 (Reptilia, Squamata): Proposed designation of a neotype. Bull. Zool. Nomencl. 48:316-318.

Myers, C. W.
    2000 A history of herpetology at the American Museum of Natural History. Bull. Amer. Mus. Nat. Hist. 252:1-232.

Myers, C. W., and J. E. Cadle
    1994 A new genus for South American snakes related to *Rhadinaea obtusa* Cope (Colubridae) and resurrection of *Taeniophallus* Cope for the *"Rhadinaea" brevirostris* group. Amer. Mus. Novitates 3102:1-33.

Myers, G. S.
    1942a Notes on Pacific Coast *Triturus*. Copeia 1942:77-82.

    1942b The black toad of Deep Springs Valley, Inyo County, California. Occ. Pap. Mus. Zool., Univ. Michigan 460:1-13.

Myers, G. S., and T. P. Maslin
    1948 The California plethodont salamander, *Aneides flavipunctatus* (Strauch), with description of a new subspecies and notes on other western *Aneides*. Proc. Biol. Soc. Washington 61:127-135.

Nesterov, P. V.
  1916 [Trois formes nouvelles d'amphibies (Urodela) du Kurdistan]. Annuaire du Musée Zoologique de L'Académie des Sciences, Petrograd 21:1-30. [In Russian.]

Ota, H., and T. Hikida
  1989 A new triploid *Hemidactylus* (Gekkonidae: Sauria) from Taiwan, with comments on morphological and karyological variation in the *H. garnotii-vietnamensis* complex. J. Herpetol. 23:50-60.

Papenfuss, T. J., J. R. Macey, and J. A. Schulte II
  2001 A new lizard species in the genus *Xantusia* from Arizona. Sci. Pap. Nat. Hist. Mus., Univ. Kansas 23:1-9.

Papenfuss, T. J., and D. B. Wake
  1987 Two new species of plethodontid salamanders (genus *Nototriton*) from Mexico. Acta Zool. Mex. (Nueva Ser.) 21:1-16.

Papenfuss, T. J., D. B. Wake, and K. Adler
  1984 (1983) Salamanders of the genus *Bolitoglossa* from the Sierra Madre del Sur of southern Mexico. J. Herpetol. 17:295-307.

Parra-Olea, G., T. J. Papenfuss, and D. B. Wake
  2001 New species of lungless salamanders of the genus *Pseudoeurycea* (Amphibia: Caudata: Plethodontidae) from Veracuz, Mexico. Sci. Pap. Nat. Hist. Mus., Univ. Kansas 20:1-9.

Pearson, O. P.
  1954 Habits of the lizard *Liolaemus multiformis multiformis* at high altitudes in southern Peru. Copeia 1954:111-116.

Pearson, O. P., and D. F. Bradford
  1976 Thermoregulation of lizards and toads at high altitudes in Peru. Copeia 1976:155-170.

Peters, J. A.
  1967 The generic allocation of the frog *Ceratophrys stolzmanni* Steindachner, with the description of a new subspecies from Ecuador. Proc. Biol. Soc. Washington 80:105-112.

Rau, C. S., and R. B. Loomis
  1977 A new species of *Urosaurus* (Reptilia, Lacertilia, Iguanidae) from Baja California, Mexico. J. Herpetol. 11:25-29.

Regal, P. J.
  1966 A new plethodontid salamander from Oaxaca, Mexico. Amer. Mus. Novitates 2266:1-8.

Resetar, A., and H. K. Voris
    1997   Herpetology at the Field Museum of Natural History, Chicago: The first one hundred years. Pp. 495-506 in Collection Building in Ichthyology and Herpetology (T. W. Pietsch and W. D. Anderson Jr., eds.). Amer. Soc. Ichthyologists Herpetologists, Lawrence, Kansas.

Richardson, C. H.
    1915   Reptiles of northwestern Nevada and adjacent territory. Proc. U. S. Natl. Mus. 48:403-435.

Riemer, W. J.
    1954   A new subspecies of the snake *Masticophis lateralis* from California. Copeia 1954:45-48.

Risch, J.-P.
    1983   *Cynops orphicus*, a new salamander from Guangdong Province, South China (Amphibia, Caudata, Salamandridae). Alytes 2:45-52.

Ritter, W. E.
    1897   The life-history and habits of the Pacific Coast newt (*Diemyctylus torosus* Esch.). Proc. California Acad. Sci., 3d Ser., Zool., 1:73-114.

    1903   Further notes on the habits of *Autodax lugubris*. Amer. Nat. 37:883-886.

Ritter, W. E., and L. Miller
    1899   A contribution to the life history of *Autodax lugubris* Hallow., a Californian salamander. Amer. Nat. 33:691-704.

Rivers, J. J.
    1889a   Description of a new turtle from the Sacramento River, belonging to the Family Trionychidae. Proc. California Acad. Sci., 2d Ser., 2:233-236.

    1889b   Habitat of *Xantusia riversiana* Cope. Amer. Nat. 23:1100.

Rodgers, T. L.
    1944   A new skink from the Sierra Nevada of California. Copeia 1944:101-104.

Rodgers, T. L., and H. S. Fitch
    1947   Variation in the skinks (Reptilia: Lacertilia) of the *skiltonianus* group. Univ. California Publ. Zool. 48:169-219.

Rodriguez-Robles, J. A., and H. W. Greeneíguez-Robles, J. A., and H. W. Greene
    1999   Food habits of the long-nosed snake (*Rhinocheilus lecontei*), a "specialist" predator? J. Zool. (Lond.) 248:489-499.

Russell, R. W., and J. D. Anderson
    1956   A disjunct population of the long-nosed salamander from the
            coast of California. Herpetologica 12:137-140. [Title should have
            read "long-toed."]

Savage, J. M.
    1952   Studies on the lizard Family Xantusiidae I. The systematic status
            of the Baja California night lizards allied to *Xantusia vigilis*, with
            the description of a new subspecies. Amer. Midl. Nat. 48:467-479.

Savage, J. M., and F. S. Cliff
    1953   A new subspecies of sidewinder, *Crotalus cerastes*, from Arizona.
            Nat. Hist. Misc. 119:1-7.

Savage, J. M., and B. I. Crother
    1989   The status of *Pliocercus* and *Urotheca* (Serpentes: Colubridae),
            with a review of included species of coral snake mimics. Zool. J.
            Linn. Soc. 95:335-362.

Schmidt, K. P., and C. M. Bogert
    1947   A new fringe-footed sand lizard from Coahuila, Mexico. Amer.
            Mus. Novitates 1339:1-9.

Seeliger, L. M.
    1945   Variation in the Pacific mud turtle. Copeia 1945:150-159.

Smith, H. M.
    1943   Another Mexican snake of the genus *Pliocercus*. J. Washington
            Acad. Sci. 33:344-345.

Smith, H. M., and D. Chiszar
    1996   Species-Group Taxa of the False Coral Snake Genus *Pliocercus*.
            Ramus Publ., Pottsville, Pennsylvania. 112 pp.

    2001   *Pliocercus bicolor*. Cat. Amer. Amphib. Rept. 737.1-737.4.

Smith, H. M., and W. W. Tanner
    1944   Description of a new snake from Mexico. Copeia 1944:131-136.

Smith, H. M., V. Wallach, and D. Chiszar
    1995   Observations on the snake genus *Pliocercus*, I. Bull. Maryland
            Herpetol. Soc. 31:204-214.

Smith, H. M., G. J. Watkins-Colwell, J. A. Lemos-Espinal, and D. Chiszar
    1997   A new subspecies of the lizard *Sceloporus scalaris* (Reptilia:
            Sauria: Phrynosomatidae) from the Sierra Madre Occidental of
            Mexico. Southwest. Nat. 42:290-301.

Song, M., X. Zeng, G. Wu, Z. Liu, and J. Fu
    2001   A new species of *Batrachuperus* from northwestern China. Asiatic Herpetol. Res. 9:6-8.

Stebbins, C. A., and R. C. Stebbins
    1953   Birds of Lassen Volcanic National Park and Vicinity. Loomis Museum Association, Mineral, California. 96 pp.

    1963   Birds of Yosemite National Park (Revised). Yosemite Natural History Association, Yosemite National Park, California. 76 pp.

Stebbins, R. C.
    1943   Adaptations in the nasal passages for sand burrowing in the saurian genus *Uma*. Amer. Nat. 77:38-52.

    1949   Speciation in salamanders of the plethodontid genus *Ensatina*. Univ. California Publ. Zool. 48:377-525.

    1951   Amphibians of Western North America. Univ. California Press, Berkeley. 539 pp.

    1954a   Amphibians and Reptiles of Western North America. McGraw-Hill, New York. 536 pp.

    1954b   Natural history of the salamanders of the plethodontid genus *Ensatina*. Univ. California Publ. Zool. 54:47-123.

    1958a   A new alligator lizard from the Panamint Mountains, Inyo County, California. Amer. Mus. Novitates 1883:1-27.

    1958b   An experimental study of the "third eye" of the tuatara. Copeia 1958:183-190.

    1960   Effects of pinealectomy in the western fence lizard *Sceloporus occidentalis*. Copeia 1960:276-283.

    1961   Body temperature studies in South African lizards. Koedoe [Journal for Scientific Research in the National Parks of the Republic of South Africa] 4:54-67.

    1966   A Field Guide to Western Reptiles and Amphibians. Houghton Mifflin, Boston. 279 pp.

    1970   The effect of parietalectomy on testicular activity and exposure to light in the desert night lizard (*Xantusia vigilis*). Copeia 1970:261-270.

Stebbins, R. C.
    1974a Off-road vehicles and the fragile desert [Part 1]. Amer. Biol. Teacher 36:203-209.

    1974b Off-road vehicles and the fragile desert [Part 2]. Amer. Biol. Teacher 36:294-304.

    1985 A Field Guide to Western Reptiles and Amphibians, 2d ed. Houghton Mifflin, Boston. 336 pp.

Stebbins, R. C., and R. E. Barwick
    1968 Radiotelemetric study of thermoregulation in a lace monitor. Copeia 1968:541-547.

Stebbins, R. C., and N. W. Cohen
    1973 The effect of parietalectomy on the thyroid and gonads in free-living western fence lizards, *Sceloporus occidentalis*. Copeia 1973:662-668.

    1995 A Natural History of Amphibians. Princeton Univ. Press, Princeton, New Jersey. 316 pp.

Stebbins, R. C., and R. M. Eakin
    1958 The role of the "third eye" in reptilian behavior. Amer. Mus. Novitates 1870:1-40.

Stebbins, R. C., and J. R. Hendrickson
    1959 Field studies of amphibians in Colombia, South America. Univ. California Publ. Zool. 56:497-540.

Stebbins, R. C., D. C. Ipsen, and G. Gillfillan
    1966 Animal Coloration: An Introduction to Natural Selection. Part I. The Principles of Concealing Coloration. Elementary School Science Project, Univ. California, Berkeley. 175 pp.

Stebbins, R. C., and C. H. Lowe Jr.
    1949 The systematic status of *Plethopsis* with a discussion of speciation in the genus *Batrachoseps*. Copeia 1949:116-129.

    1951 Subspecific differentiation in the Olympic salamander *Rhyacotriton olympicus*. Univ. California Publ. Zool. 50:465-482.

Stebbins, R. C., J. M. Lowenstein, and N. W. Cohen
    1967 A field study of the lava lizard (*Tropidurus albemarlensis*) in the Galápagos Islands. Ecology 48:839-851.

Stebbins, R. C., and W. J. Riemer
  1950   A new species of plethodontid salamander from the Jemez Mountains of New Mexico. Copeia 1950:73-80.

Stebbins, R. C., W. Steyn, and C. Peers
  1960   Results of stirnorganectomy in tadpoles of the African ranid frog, *Pyxicephalus delalandi*. Herpetologica 16:261-275.

Stein, B. R.
  2001   On Her Own Terms: Annie Montague Alexander and the Rise of Science in the American West. Univ. California Press, Berkeley. 380 pp.

Stickel, W. H.
  1938   The snakes of the genus *Sonora* in the United States and Lower California. Copeia 1938:182-190.

Storer, T. I.
  1922   The California Museum of Vertebrate Zoology. California Alumni Monthly 15:257-259.

  1925   A synopsis of the Amphibia of California. Univ. California Publ. Zool. 27:1-343.

  1929   Notes on the genus *Ensatina* in California, with description of a new species from the Sierra Nevada. Univ. California Publ. Zool. 30:443-452.

  1959   Some Pacific Coast zoological history. Bios 30:131-147.

Tanner, V. M.
  1951   Pacific Islands herpetology, No. V Guadalcanal, Solomon Islands: A check list of species. Great Basin Nat. 11:53-86.

Tanner, W. W.
  1981   A new *Hypsiglena* from Tiburon Island, Sonora, Mexico. Great Basin Nat. 41:139-142.

  1985   Snakes of western Chihuahua. Great Basin Nat. 45:615-676.

Twitty, V. C.
  1935   Two new species of *Triturus* from California. Copeia 1935:73-80.

Wake, D. B.
  1978   Tracy I. Storer and Charles L. Camp. Copeia 1978:196-197.

  1991   Homoplasy: The result of natural selection, or evidence of design limitations? Amer. Nat. 138:543-567.

Wake, D. B.
1996a A new species of *Batrachoseps* (Amphibia: Plethodontidae) from the San Gabriel Mountains, southern California. Contrib. Sci., Nat. Hist. Mus. Los Angeles County 463:1-12.

1996b Evolutionary developmental biology—Prospects for an evolutionary synthesis at the developmental level. Pp. 97-107 in New Perspectives on the History of Life: Essays on Systematic Biology as Historical Narrative (M. T. Ghiselin and G. Pinna, eds.). Memoirs of the California Academy of Sciences, No. 20. California Acad. Sci., San Francisco.

1997 Incipient species formation in salamanders of the *Ensatina* complex. Proc. Natl. Acad. Sci. USA 94:7761-7767.

Wake, D. B., and A. H. Brame Jr.
1969 Systematics and evolution of Neotropical salamanders of the *Bolitoglossa helmrichi* group. Contrib. Sci., Nat. Hist. Mus. Los Angeles County 175:1-40.

Wake, D. B., A. H. Brame Jr. II, and R. Thomas
1982 A remarkable new species of salamander allied to *Bolitoglossa altamazonica* (Plethodontidae) from southern Peru. Occ. Pap. Mus. Zool., Louisiana State Univ. 58:1-21.

Wake, D. B., and J. A. Campbell
2001 An aquatic plethodontid salamander from Oaxaca, Mexico. Herpetologica 57:509-514.

Wake, D. B., and P. Elias
1983 New genera and a new species of Central American salamanders, with a review of the tropical genera (Amphibia, Caudata, Plethodontidae). Contrib. Sci., Nat. Hist. Mus. Los Angeles County 345:1-19.

Wake, D. B., and J. D. Johnson
1989 A new genus and species of plethodontid salamander from Chiapas, Mexico. Contrib. Sci., Nat. Hist. Mus. Los Angeles County 411:1-10.

Wake, D. B., and J. F. Lynch
1976 The distribution, ecology, and evolutionary history of plethodontid salamanders in tropical America. Sci. Bull. Nat. Hist. Mus. Los Angeles County 25:1-65.

Wake, D. B., and J. F. Lynch
    1982   Evolutionary relationships among Central American salamanders of the *Bolitoglossa franklini* group, with a description of a new species from Guatemala. Herpetologica 38:257-272.

Webb, R. G.
    1975   Taxonomic status of *Aspidonectes californiana* Rivers, 1889 (Testudines, Trionychidae). Copeia 1975:771-773.

Wilson, L. D.
    1982   A review of the colubrid snakes of the genus *Tantilla* of Central America. Contrib. Biol. Geol., Milwaukee Public Mus. 52:1-77.

    1983   A new species of *Tantilla* of the *taeniata* group from Chiapas, Mexico. J. Herpetol. 17:54-59.

Wilson, L. D., J. R. McCranie, and K. L. Williams
    1985   Two new species of fringe-limbed hylid frogs from nuclear Middle America. Herpetologica 41:141-150.

Wood, W. F.
    1940   A new race of salamander, *Ensatina eschscholtzii picta*, from northern California and southern Oregon. Univ. California Publ. Zool. 42:425-427.

Zhao, E.
    1995   Infraspecific classification of some Chinese snakes. Sichuan J. Zool. 14:107-112. [In Chinese.]

Zweifel, R. G.
    1952   Pattern variation and evolution of the mountain kingsnake, *Lampropeltis zonata*. Copeia 1952:152-168.

    1954a   A new frog of the genus *Rana* from western Mexico with a key to the Mexican species of the genus. Bull. Southern California Acad. Sci. 53:131-141.

    1954b   A new species of *Chersodromus* from Mexico. Herpetologica 10:17-19.

    1955   Ecology, distribution, and systematics of frogs of the *Rana boylei* group. Univ. California Publ. Zool. 54:207-291.

    1959   Variation in and distribution of lizards of western Mexico related to *Cnemidophorus sacki*. Bull. Amer. Mus. Nat. Hist. 117:57-116.

Zweifel, R. G., and K. S. Norris

1955   Contribution to the herpetology of Sonora, Mexico: Descriptions of new subspecies of snakes (*Micruroides euryxanthus* and *Lampropeltis getulus*) and miscellaneous collecting notes. Amer. Midl. Nat. 54:230-249.

www.ingramcontent.com/pod-product-compliance
Lightning Source LLC
Chambersburg PA
CBHW080426270326
41929CB00018B/3175